SUCCESS
STRATEGIES
FOR WOMEN
IN SCIENCE

· · · · · · · · · · · · · · · ·

A PORTABLE MENTOR

Edited by
Peggy A. Pritchard

ELSEVIER

AMSTERDAM • BOSTON • HEIDELBERG • LONDON
NEW YORK • OXFORD • PARIS • SAN DIEGO
SAN FRANCISCO • SINGAPORE • SYDNEY • TOKYO

Academic Press is an imprint of Elsevier

Elsevier Academic Press

30 Corporate Drive, Suite 400, Burlington, MA 01803, USA
525 B Street, Suite 1900, San Diego, California 92101-4495, USA
84 Theobald's Road, London WC1X 8RR, UK

This book is printed on acid-free paper. ∞

Library of Congress Cataloging-in-Publication Data
An application has been submitted

British Library Cataloguing in Publication Data
A catalogue record for this book is available from the British Library

ISBN 13: 978-0-12-088411-7
ISBN 10: 0-12-088411-9

For all information on all Elsevier Academic Press publications
visit our Web site at www.books.elsevier.com

Printed in the United States of America
05 06 07 08 09 10 9 8 7 6 5 4 3 2 1

Working together to grow
libraries in developing countries

www.elsevier.com | www.bookaid.org | www.sabre.org

ELSEVIER   BOOK AID
             International   Sabre Foundation

SUCCESS
STRATEGIES
FOR WOMEN
IN SCIENCE

· · · · · · · · · ·

A PORTABLE MENTOR

This book is dedicated to my grandmothers:

F.G. "Trudy" McLean (nee Adams)
M.E. "Peggy" Pritchard (nee Nelson)

and my granddaughters:

Olivia Mary-Rose Blackwell
Allison Sara Kropinski

Keep me away from the wisdom that does not cry, the philosophy that does not laugh and the greatness that does not bow before children.[1]

Your children are not your children.
They are the sons and daughters of Life's longing for itself.[2]

Kahlil Gibran, Lebanese Poet

---

[1] From Kahlil Gibran, "Mirrors of the Soul" [in] Mirrors of the Soul, translated and with biographical notes by Joseph Sheban (New York: Philosophical Library, 1965:72)

[2] Kahlil Gibran, "On Children" [in] The Prophet (NY: Knopf, 1923:17).

CONTENTS

CONTRIBUTORS

Information in parentheses indicates the chapter(s) to which the author contributed.

MARGARET-ANN ARMOUR (12), Department of Chemistry, University of Alberta, Edmonton, Alberta T6G 2G2 Canada

MARK BISBY (Foreword), Canadian Institutes of Health Research, Ottawa, Ontario K1A 0W9 Canada

L. CATE BRINSON (11), Department of Mechanical Engineering, Northwestern University, Evanston, Illinois 60208 United States

ILENE J. BUSCH-VISHNIAC (4), Department of Mechanical Engineering, Johns Hopkins University, Baltimore, Maryland 21218 United States

M. ELIZABETH CANNON (8), Department of Geomatics Engineering, University of Calgary, Calgary, Alberta T2N 1N4 Canada

VALERIE ANN CORNISH (11), Biochemist, Durham DH1 4NE United Kingdom

CHRISTINE FAERBER (2), Competence Consulting, Potsdam D-14480 Germany

CHRISTINE S. GRANT (5), Department of Chemical Engineering, North Carolina State University, Raleigh, North Carolina 27695 United States

ANNGIENETTA JOHNSON (2), National Aeronautics and Space Administration, Washington, D.C. 20546 United States

Joyce McCarl Nielsen (6), College of Arts and Sciences, University of Colorado at Boulder, Boulder, Colorado 80309 United States

Mary Osborn (3), Department of Biochemistry and Cell Biology, Max Planck Institute for Biophysical Chemistry, Goettingen 37077 Germany

Peggy A. Pritchard (Introduction, 1, 7, 10), Pritchard Communications and Consulting, Guelph, Ontario K1G 2L8 Canada

Sarah E. Randolph (11), Department of Zoology, University of Oxford, Oxford OX1 3PS United Kingdom

Patricia Rankin (6), College of Arts and Sciences, University of Colorado at Boulder, Boulder, Colorado 80309 United States

Nadia Rosenthal (Prologue), European Molecular Biology Laboratory, Monterotondo, Rome I-00016 Italy

Linda S. Schadler (11), Department of Materials Science and Engineering, Rensselaer Polytechnic Institute, Troy, New York 12180 United States

Kathleen Sendall (8), North American Gas, Petro-Canada, Calgary, Alberta T2P 3E3 Canada

Miriam Stewart (Foreword), Institute of Gender and Health, Canadian Institutes of Health Research, University of Alberta, Edmonton, Alberta T6G 2T4 Canada

Christine Szymanski (9), Institute for Biological Sciences, National Research Council of Canada, Ottawa, Ontario K1A 0R6 Canada

Dorothy Tovell (12), Biochemist, Edmonton, Alberta T6G 0A7 Canada

FOREWORD

········
················

When we were approached with the innovative idea for this timely book, we were pleased to provide modest funding from the Canadian Institutes of Health Research (CIHR) to assist the Editor in making contacts, travelling, and conducting interviews that have led to many of the informative and inspiring contributions. CIHR did this because of its belief that women and men should both be able to pursue careers in research, limited only by their talent and ambition. Any institutional or cultural factors that discourage their participation should be identified and removed. However, changing institutions and cultures is challenging and slow. So, while we pursue this long-term goal as co-chairs of an inter-agency group comprised of ten national research funding agencies and academic organizations, *Success Strategies for Women in Science: A Portable Mentor* provides welcome and much-needed guidance to women contemplating and coping with the status quo. Indeed, there is advice and encouragement here for everyone.

We willingly agreed to write this Foreword, but recognize that, within our own sphere of influence, there is much to be done. CIHR's own record must be improved. To take any comfort in the fact that our experience seems typical of many national funding agencies is to fail to acknowledge the pervasive problem of missing women. At every career transition in health research in Canada, women disappear (Fig. 1).

There is no clear discrimination against women in the adjudication process for our grants and awards: women who apply do as well as men in our competitions for funding. The problem is that women choose not to compete. In our 2003 competition for Ph.D. studentships, 58% of the applicants were women; for postdoctoral fellowships, 47% were women; and for career awards for new faculty, only 36% were women. When we last offered

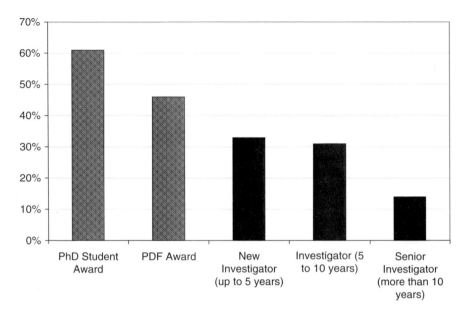

Figure 1   Proportion of CIHR Training (grey) and Career Awards (black) held by Women, 2004–2005.

awards for senior investigators (1992), women comprised only 20% of award holders. Why is health research losing this talent at successive career stages, and what can we do about it?

On the face of it, the situation seems to be improving quite rapidly. Participation of women in the national Ph.D. award competition in health research increased from 46% to 58% between 1997 and 2003, with comparable gains for the postdoctoral and new investigator awards over this period (Table 1).

However, the apparent gains during this time are primarily explained by the creation, in 2000, of CIHR, an agency with a much broader mandate than its predecessor (which focussed primarily on biomedical and clinical research). CIHR's mandate encompasses health services and population health research, including research on the social, cultural, and environ-

Table 1:   Percentage of women applicants to awards competitions of the Canadian federal health research funding agency

	Ph.D. award	Postdoctoral	New investigator
1997	46%	35%	28%
2003	58%	47%	36%

Table 2: Percentage of women applicants (and number) to CIHR's 2003 awards competitions, in four theme areas of health research

	Biomedical	Clinical	Health services	Population health
PhD award	46% (248)	73% (91)	69% (21)	83% (64)
Postdoctoral	40% (93)	56% (34)	67% (18)	81% (17)
New investigator	25% (52)	48% (30)	53% (18)	50% (31)

mental determinants of health. The influx of applications from these more social science-oriented fields of health research accounts for most of the increased participation of women. For example, at the Ph.D. level, 69% of applications in the area of health services research were from women, while in population health, women were in the clear majority: 83% of applicants (Table 2). There was similar, strong representation of women in these areas at the postdoctoral and new investigator award level. Unfortunately, participation in the biomedical sciences (by far the largest number of applicants) was little improved from the 1997 levels (compare Table 1, top row, with Table 2, column 2). The progressive loss of representation of women at the different career stages is also greatest for the biomedical and clinical areas, and much less for health services and population health research.

This remarkable difference, between the biomedical and clinical sciences and the areas of health services and population health, is also marked at more senior career stages. CIHR suspended its senior career awards programs following the 2002 competition, but in 2001 it held separate competitions in the biomedical/clinical and health services/population health areas for Investigator Awards (for those with more than five years of independent research experience), and Senior Investigator Awards (more than 10 years experience). The participation of women was dramatically different in the two competitions (Table 3).

Table 3: Percentage of women applicants (and number) to CIHR's 2001 senior career awards competitions, held separately for biomedical/clinical and health services/population health streams.

	Biomedical/Clinical	Health Services/ Population Health
Investigator	29% (26)	65% (11)
Senior Investigator	19% (12)	44% (8)

These data raise some interesting questions. What is it about population health and health services research that attracts women, and what deters women from contemplating a career in academic biomedical research? The competition for funding is not any less severe in the areas of health services and population health research; indeed, the severity of reviewer criticism, and negativity of reviews, is actually worse in this area than in biomedical research (Thorngate et al., 2002). If women participate more than men in some areas of health research, why should we be concerned that they are underrepresented in others? The reduced proportion of women applying for more senior awards in the biomedical and clinical areas might be due to the greater availability of alternate career paths outside academia for those with research training in these areas of health research, and surely this is not a bad thing at all. The question, of course, is the extent to which the women who disappeared from the academic career track did so because they really wanted to, or because they were deterred from their first choice.

The fact remains that in the largest area of health research, biomedical research, our Canadian experience is that women are seriously underrepresented in academia, and the situation is not improving very quickly. Further worrying evidence is that, in the health sciences, the proportion of women nominated by Canadian universities for the prestigious, "Tier 1 Canada Research Chairs," has been about half the proportion of women eligible for nomination (Begin-Heick, 2001). Moreover, the gap between eligibility and nominations for Canada Research Chairs is actually greater for the health sciences than for the physical sciences and engineering, those traditional bastions of male domination.

We are encouraged that senior representatives of key research funding organizations in Canada are collaborating by sharing strategies that foster success and sustainability in women's research careers. The two challenges they identified (that are prevalent across health-related research organizations in Canada) are: 1) the diminished participation of women at more senior levels, and 2) the higher participation of women in more social-cultural sciences than in basic and natural sciences. Moreover, these challenges were the key themes of an international research gathering in Washington in the fall of 2003: *A Colloquium on Career Paths for Women in the Life Sciences: A Global Perspective*. Participants included more than 70 academics, administrators and funding agency representatives. Recommendations for action included identification of resources to support gathering of country-specific data, determination of strategies to use the Internet and other electronic means to mentor and support women in the life sciences in the developing world, and determination of priority political venues that could be accessed to work toward enhancement of career options for women in the life sciences in low- and middle-income nations. Participants at this event were impressed by collaborations in Canada designed to foster women's research careers.

It will take time to change the systems in which scientific research is undertaken, even in a country that prides itself on the progress it has made in advancing gender equality. In the meantime, this book will help you understand what *you* can do to enhance your success, inspire you to take advantage of opportunities as they arise (or create them yourself), and encourage you to work within the system to make your own contribution to change. You will find chapters dealing with essential skills required of every researcher, such as: networking, communicating, coping with the demands of a research career, time management, and the most difficult of skills, saying "no" to excessive demands on your time. Issues relating to career development are explored, and the importance of the examination of alternate career paths is stressed. While much of the advice in this mentoring manual is aimed at women beginning their careers, readers at other stages will find the book of value. Staying current in a narrowly specialized field, while maintaining an innovative and comprehensive approach to complex research problems, are difficult intellectual challenges, regardless of your level. And we all struggle to balance the time and energy we devote to our two loves: our work, and our family and friends.

While the individual strategies in this book are important, we want to stress again that we recognize that significant systemic changes are also needed.

"The opportunity to do science every day is about as much fun as you can have over a lifetime," reported one respondent in a survey of over 6,000 life scientists conducted by the American Association for the Advancement of Science (Holden, 2004:1830). And, for the most part, these scientists—women and men—were happy with their career; only 8% regretted their choice. However, as another respondent pointed out "this is a survivor survey." The need for better career counseling was one of the main trends identified in the survey. This book is one step in the right direction. And while a scientific career is intellectually rewarding like few others, there will be times when you desperately need advice and mentorship but won't be able to get what you need from friends or colleagues. We hope this book will fill the gap, and help women to pursue excellence and achieve success in their chosen, scientific careers.

Miriam Stewart
Scientific Director
Institute of Gender and Health
Canadian Institutes of Health Research

Mark Bisby
Vice-President
Research Portfolio
Canadian Institutes of Health Research

REFERENCES

Begin-Heick, N. and Associates. *Gender-based analysis of the Canada Research Chairs Program: A report prepared for the Canada Research Chairs Secretariat.* (2002) [Online]. Available from the Internet: www.chairs.gc.ca/web/about/publications/gender_e.pdf (accessed February14, 2005).

Holden, C. Long hours aside, respondents say jobs offer "as much fun as you can have." *Science.* 2004;18(304): 1830–1837.

Thorngate, W., et al. *Mining the archives: Analyses of CIHR research grant adjudications* (2002) [Online]. Available from the Internet: www.carleton.ca/~warrent/reports/mining_the_archives.pdf (accessed February14, 2005).

INTRODUCTION

INSPIRATION

In 2000 I was appointed to a department of microbiology at a research-intensive university in Canada to run a novel, graduate-level, professional skills training course that I had developed with a colleague. Though I had advised and mentored graduate students for years, it was through this course that I observed, firsthand, the benefits of mentoring to aspiring scientists. Not only did my students learn from a trans-disciplinary team of knowledgeable experts and award-winning instructors, they enjoyed privileged access to people who took a personal interest in them and their careers. This involvement inspired them to do their best.

My own experience with the students was no exception. As our relationship grew and they felt more confident sharing their concerns and questions about their futures, they became more active in finding answers for themselves, and more confident in their choices and in the directions they were taking.

They asked many excellent questions. But I was not always the best person to answer because my expertise and career goals were not a good match with theirs. I encouraged them to seek mentors on their own, but often, they were reluctant to do so. Much of their reticence stemmed from the difference in status and power between them and the successful scientists they had identified as being the best people to mentor them. They were convinced that these successful people were unapproachable, would not be interested in sharing their insights, and would want nothing to do with a young scientist and her questions.

I knew otherwise, but could not convince them. So I thought: "If my students won't go to the mentors, I'll bring the mentors to the students." And thus, the idea for *Success Strategies for Women in Science: A Portable Mentor* was born. Thanks to seed funding from the Canadian Institutes of Health Research, I was able to interview over 350 scientists across North America and Europe, of whom 18 became major contributors to this work. Many more are represented in the stories recounted in each chapter.

The Problem

In North America and Europe, senior administrators and policy makers in government, research institutes, industry and academe, as well as employers, members of science associations, and educators, have long recognized the need to actively encourage women to enter and stay in the sciences. Attrition rates in secondary schools, higher education, and the early stage of scientific careers are well documented, as is the under representation of female scientists in senior positions in all sectors of scientific endeavor. While many organizations are actively examining and addressing the systemic problems, "changing institutions and cultures is challenging and slow."[1]

In the meantime, women contemplating work in science, and those already well on their way, need to be able to function (and succeed) within the systems that exist *now*, flawed as they are. But how can this best be achieved? Working groups investigating the obstacles to the advancement of women in science have identified "mentorship" as *the* most important factor contributing to the success of female scientists. Though mentoring programs for women do exist and take various forms, access is not universal. This book will complement these efforts and support women whose mentoring opportunities are limited.

Purpose

Success Strategies for Women in Science: A Portable Mentor distills into one volume a wealth of knowledge and years of experience of success-

---

[1] Drs. Mark Bibsy, Vice President, Research Portfolio, Canadian Institutes of Health Research (CIHR), and Miriam Stewart, Director of the Institute of Gender and Health, CIHR, in the Foreword to this book.

ful female scientists from industry, government, research institutes, and academe. Through practical advice and real-life stories, readers will learn what knowledge and skills are needed to make the transition from trainee to scientist that, if practiced, will help them to become successful. The book addresses *current* issues and concerns these women will face as they plan their careers.

This is *not* to suggest that the "System" is perfect, or that there is something wrong with women, but rather that by focusing on what young scientists can do to *help themselves*, they may experience a sense of empowerment that will inspire behaviors linked to success in science. Of course, this is only *one* aspect of the solution to the problem of the under-representation of women in science, but an important one over which readers have control.

OUTSTANDING CONTRIBUTORS

The strength of the book derives from the experience, stature, and success of the contributors and colleagues whose stories they share, and their personal commitment to promoting and supporting women in science. They represent diverse scientific backgrounds and interests, including the life and human sciences, physical and social sciences, and pure and applied sciences. Many have written extensively on the issue of the under-representation of women in science; most are actively involved in *doing* something about it, at all levels, from working directly with trainees in mentorship programs to making institutional and government policy more equitable.

SCOPE

Throughout this work, the term "science" is used very broadly to mean the organized body of knowledge gained through a process of systematic study involving observation and experimentation. Traditionally, the sciences are grouped under headings such as the physical sciences (e.g., chemistry, physics), life sciences (e.g., botany, molecular biology), applied sciences (e.g., engineering, mathematics, computing) and the social sciences (e.g., archaeology, criminology). These artificially-constructed categories are by no means mutually exclusive; rather, as our knowledge and understanding expand, new relationships emerge and with them, fields of study (e.g. sociogenomics, bioinformatics and paleoarchaeology). Some chapters in this book make the distinction between engineering and the other sciences, but generally, the term is used to include all disciplines.

It is assumed that readers, in their drive to succeed in science, already are applying themselves to developing a thorough understanding of the literature in their specific disciplines (and related disciplines, as appropriate), excellent bench skills, and other competencies directly related to mastery in their chosen fields. This book does not address itself to these. Rather, the twelve topics covered in this book are those related to skills and knowledge that *complement* scientific training and expertise; that will *enhance* the potential for success. They are:

Career Management
Continuing Professional Development
Training and Working Abroad
Climbing the Ladder
Mentoring
Networking
Mental Toughness
Personal Style
Communicating Science
Time Stress
Balancing Professional and Personal Life
Transitions

INTENDED AUDIENCE

Trainees and Early Career Scientists

This manual was developed for women nearing the end of their formal training and beginning their careers in science, to be a "portable" mentor. But the topics are not specific to women's success only. Everyone interested in taking an active role in their own career, in doing all they can to achieve success, will find advice, support, and encouragement. Male—as well as female—graduate students, postdoctoral fellows, and early career scientists (particularly those from under-represented groups) will benefit from the strategies suggested by role models who are passionate about helping them succeed. Students finishing their undergraduate degrees and young women in their senior years of secondary school who are considering careers in science, will also be inspired to "aim for the stars."[2]

---

[2] Dr. Shirley Ann Jackson, President of Rensselaer Polytechnic Institute, received this advice from her father when she was a young girl. It has been an inspiration all her life.

Supervisors, Mentors, Advisors, Parents

Mid-career and senior scientists (of both sexes) involved in mentoring female graduate students, doctoral fellows, and younger colleagues, will gain insight from these exemplars on how to become better mentors, and will have a practical resource to recommend to them. This is also true of career counselors in colleges and universities, guidance counselors and science teachers in public and private secondary schools, and parents, who are involved in supporting young women as they consider their next steps for education and work.

CONTINUING THE CONVERSATION

Success Strategies for Women in Science: A Portable Mentor is intended as a practical resource to address the immediate (and continuing) need to bring together the community of scientists in an atmosphere of openness and mutual trust, for the exchange of ideas, experiences, and encouragement. My hope is that the conversations begun in this book will continue among readers—in their own contexts, with their own colleagues—and expand to include new voices, in an increasingly diverse society.

Peggy A. Pritchard

ACKNOWLEDGMENTS

This mentoring manual would not exist if it weren't for the more than 350 wonderful women scientists I interviewed in Europe and North America—undergraduate students to senior research scientists and emerita professors—whose stories breathed life into the work; my graduate students, whose tentative questions about their futures in science provided the first inspiration for the project; and especially, my Contributors who, in spite of their already too-full schedules, were so willing to share their enormous wisdom, experience, humor, and enthusiasm by contributing chapters. I hope that the virtual meeting of hearts and minds presented in this book will become *real* in the life of each reader.

In particular, I wish to thank:

Dr. Ursula Franklin, Physicist, Professor Emerita and Companion of the Order of Canada, for her incisive mind, probing questions, and for making me aware of the broader and deeper issues.

Dr. Rita Colwell, Chairman, Canon U.S. Life Sciences, Inc. and Past Director of the National Science Foundation, whose early support of the project gave it credibility and inspired me never to give up.

Dr. Shirley Ann Jackson, President of Rensselaer Polytechnic Institute, whose commitment to "aim for the stars" is a model for others to make the most of their own potential.

Dr. Mark Bisby, Vice President, Research Portfolio, Canadian Institutes of Health Research (CIHR), whose patience and determination in the pursuit of excellence and the advancement of women in science are qualities to which I aspire.

Dr. Alan Bernstein, President of CIHR, and Dr. Miriam Stewart, Director of the Institute of Gender and Health, CIHR, for providing seed funding that enabled me to travel.

Dr. Valerie Davidson, National Sciences and Engineering Research Council/Hewlett-Packard Chair for Women in Science and Engineering, and the School of Engineering, University of Guelph, Canada, for their generous donation of office space, and their tremendous welcome and collegial support during my time as Visiting Scholar.

My editors at Elsevier, Ms. Lisa Tickner, whose easy communication style and collaborative approach to our working relationship have made the entire process a delight, and Ms. Clare Rathbone, whose gentle prodding at the end helped me close the book.

To my network of colleagues, friends and family: words cannot express my sincere appreciation for your ongoing support and encouragement, which sustained me throughout the project. Dr. Sally Ann Amero, Scientific Review Administrator, National Institutes of Health Center for Scientific Review, who sat with me one afternoon in the summer of 2000 and brainstormed about a book for women in science; Dr. Peter Aston, former Head of the Department of Microbiology and Immunology, Queen's University, Canada, for formally recognizing my contributions to graduate teaching in the department and the Faculty of Health Sciences; and my many reviewers, including Drs. Kathy Barker, Marie-Luce Constant, Irene Karsten, Linda Hawkins, and Susan Read.

Finally, to my parents, Dr. Ruggles B. Pritchard and Mrs. Elizabeth R. Pritchard, who volunteered to inconvenience themselves for an extended period of time by babysitting a bossy Basenji; and to Dr. Andrew M. Kropinski, Senior Scientist, Public Health Agency of Canada, my staunchest supporter, life partner, mentor, friend, gourmet chef, and "Internet Bloodhound Extraordinaire": Thank You.

NADIA ROSENTHAL, Ph.D.
Coordinator, Monterotondo Research Programme
European Molecular Biology Laboratory, ITALY
Developmental Biologist

• •

For Dr. Nadia Rosenthal, life in science is a great adventure, filled with riddles, possibilities, and surprising interrelationships. The face of the unknown—which can inspire insecurity and hesitancy in some—*fuels* her passion for discovery. Her infectious enthusiasm and excitement for research, and commitment to supporting and encouraging others, inspire all who meet her.

A leading researcher in muscle gene regulation, Dr. Rosenthal completed her doctorate in biochemistry at Harvard Medical School and postdoctoral fellowship in molecular virology at the National Cancer Institute (NCI). She was a Staff Fellow at NCI before moving on to faculty positions at Boston University School of Medicine and Harvard Medical School. As she pursued her interests and developed her research program, she remained open to serendipitous lines of inquiry and creative approaches to funding that resulted in the development of a strong, successful research program in muscle cell development and disease.

In 2001 she was lured away from Harvard to head the Mouse Biology Programme at the European Molecular Biology Laboratory (EMBL) in Monterotondo, Italy (near Rome), one of Europe's premier research institutes. The first American and one of the first women to head an EMBL center, she has built the programme from 12 to 80, and currently runs a laboratory of 10 researchers.

Throughout her career, her commitment to research, training, and the advancement of science have expressed themselves through her work and administrative activities—at institutional, regional, national, and international levels. Currently, she is heavily involved in policy issues around Europe, as a member of the European Group on Life Sciences.

Dr. Rosenthal would be the first to acknowledge that Science is a challenging field—for men as well as for women. But she believes its rewards are worth the effort. "It's a great life," she says. "Discovery is the best 'high' there is!"

PROLOGUE

·················

A MAGNIFICENT OBSESSION

The master painter disposes the colours for the sake of a picture that cannot be seen in the colours themselves.

-The Buddha

I never thought of myself as a scientist, at least not in the way scientists are conventionally portrayed to the public: solitary, disheveled figures working late, bending over bubbling beakers, with calculators in the pockets of their lab coats, oblivious to their surroundings. Of course that probably is exactly what I looked like as a molecular biology graduate student at Harvard in the 1970s, but inside my head I was exploring a world most people never have a chance to see. I was a naturalist of the nucleus, on a trail of detection that was as exciting as anything I had ever encountered. My childhood in a family of artists had prepared me for a different obsession, but this new world, opened up by an inspiring high school teacher, was even more compelling. She showed us how awesome was nature in its detail, beautiful and unpredictable. And I was hardly solitary. I felt I was swimming in a broad stream with all the other biologists who had worked before me and the ones who will come after. The history of science is not a history of humans, but of human discoveries, measured not against each other, but against nature itself. That was what gave me strength during the times when the going got rough later on. The promise of a truth that would stand up to Nature's scrutiny made the hard work and endless obstacles of no particular consequence to me.

Decades later, I still feel this way, and I do whatever I can to foster the same excitement in my own students. Many are women, I am glad to say, and to date, not a single one of these wonderful scientists has been lost to the profession. There is safety in numbers, and the numbers are growing. When I was awarded my Ph.D., I was the only female student left from my entering class. We had started out well—an equally balanced group—but the intervening years had taken their toll; although it was a personal triumph to have survived the process, it also reinforced my fear of future failure. One more casualty that year, and there would not have been a single woman on the podium at graduation.

As scientists, we are all bitten by the same bug of universal curiosity and have the same dread of personal failure, but women have the additional burden of discrimination. At the beginning of the 21st century, we are generally comfortable with the abstract notion that a woman is equally entitled to satisfy her scientific curiosity. Female life scientists abound in academic institutions, at least until the positions and money and space become limited (usually at the Associate Professorship level), then the attrition rate is embarrassingly high. And it's not just about children. There are plenty of childless women on the dropout list, and those who have attained positions of power in their profession are just as likely to have children as not. There are a thousand subtle and not-so-subtle ways to discourage a young researcher, to distract her from the joys of discovery and dissuade her from demanding more space or more support when she clearly needs and deserves it.

It's important to identify our own impediments. We are not all well enough equipped to deal with competition—for positions, promotions, or papers—and competition is a constant in research. Above all, we need to recognize the power imparted by external research funding. I once sat on a committee to analyze the plight of women (or lack thereof) in senior research positions at my institute and was mortified to discover that female laboratory heads were receiving, on average, 40% less funding from the National Institutes of Health (NIH) than their male counterparts. For years I had served on NIH grant review panels; how could I have missed this blatant discrimination on the part of my colleagues? The real horror struck when we examined the data in detail: the women had asked for 40% less money on their applications. I repeat this tale to every young postdoctoral fellow leaving my laboratory. They understand the message: "Male or female, you won't get what you do not ask for in this world."

How then, do we promote a sense of entitlement amongst women in science? How can we protect the original obsession that drives us into the field in the first instance, and fires the necessary engines to steer one's personal path through the obstacle course of today's competitive research environment? It's a multifaceted problem that requires much more attention than it has received. This book is a rich and varied resource of insights from leading women scientists on aspects of training, mentoring, networking, and communication, so essential to a successful scientific career. Each of these women has engaged her curiosity in diverse and marvelous ways. I would argue that the common denominator among these powerful leaders has been a personal passion for science, of which they never lost sight.

Though it hits everyone differently, the feeling of infectious curiosity is unmistakable. My own obsession with science sprang, unexpectedly, out of an early passion for art. I began drawing as soon as I could hold a pencil. I'd draw anything I could. I wanted to see if I could get it just "right"—by high school I had gone to the extremes of hyper-realistic painting. My epiphany at fifteen was sparked as much by the recurring themes in nature that I had been trying to capture in paint as by the phylogeny, evolutionary biology, and biochemical pathways of metabolism I was gobbling up at school. The spirals in seashells and sunflower heads and the shifting symmetries of embryonic body plans raised persistent questions about general form and the forces that shape them, and convinced me that the biology of pattern formation would satisfy my curiosity more than painting ever could. In my ignorance, I was sure that the processes of developmental biology had been worked out to the same degree of mechanistic detail as on my intermediary metabolic charts, and that at university, Nature would reveal her morphological secrets to me.

Of course, I never found those morphogenetic charts at university, nor anywhere else for that matter. It wasn't until I came across a popular science magazine in my university library that I realized how limited the collective knowledge was at that time. A picture of a child's outstretched hand was on the cover. The caption read: "How Does a Hand Know to Become a Hand?" but the article didn't shed much light on the actual process of limb patterning, and instead posited the presence of hypothetical morphogen gradients and reviewed current concepts of positional information. I was fascinated, and rushed off to do more reading, but emerged disappointed by the lack of mechanistic detail in the articles I found. My professors only affirmed what I suspected: the field was awaiting the molecular revolution that would take another two decades to unfold.

In the interim, I found other satisfactions: first in the revelation of evolution at work as we caught our first glimpses of mammalian gene structure; then in the pursuit of elusive molecular interactions underlying the new genetic code of eukaryotic gene regulation; later in the excitement of testing our hypotheses of transcriptional control in living animals through transgenic and gene knockout technologies. It has been a capricious path, but peopled with marvelous colleagues, and the synthesis of collective discovery is a joy for which nothing I learned from my textbooks could have prepared me. Despite the practical difficulties and psychic pitfalls, I have maintained a sense of freedom to pursue my curiosity—not only because of some lucky breaks along the way, but because I found I just couldn't put up with anything less. I tell my students to do the same when they enter the laboratory, and it has paid off over and again. Thirty-five years after my original epiphany, I am finally returning to the problem of vertebrate limb morphology, thanks to a brilliant student who showed me how to approach the subject in a novel way, using all the wondrous tricks of the trade we now have at our disposal.

As I reflect on the characteristics that help scientists realize their dreams, I am impressed by the resilience we need to withstand the tribulations of the profession in order to keep focused on discovery and on the promise of epiphany that originally drew us into the field. Any strategies we develop or employ to survive and flourish must begin with seizing the moment as it unfolds and using it to our best advantage. Patience is not the virtue I would espouse here, but rather a stubborn intolerance of personal compromise when it comes to pursuing your ideas. It takes clever strategizing to keep doing what you're interested in doing, in the face of shifting fashions and inconsistent funding. The politics and practicalities of research are necessary parts of the game, and can work just as well *in* your favor as against it. But the centrepiece has to be the science. If you are truly obsessed with a magnificent question, Nature never lets you forget it.

Nadia Rosenthal, Ph.D.
Head, EMBL Mouse Biology Programme
European Molecular Biology Laboratory
Monterotondo (Rome)
Italy

Chapter

1

• • • • • • • • • •

CAREER MANAGEMENT

• • • • • • • • • • • • • • • • • •

Peggy A. Pritchard, Editor

> *I always had the next desirable step in view, but never the whole thing. However, my professional life came out far better than I ever could have planned as a young woman.*
>
> Joanne Simpson, Chief Scientist for Meteorology, NASA

THE MYTH OF "CAREER PLANNING"

Most of the 350+ female scientists interviewed for this book reported that they did not "plan" their careers, at least not in the traditional sense of mapping out every step that they would take, from the time of their formal training, to their retirement, and beyond. Rather, at each stage of their professional and personal lives, they looked ahead to the next transition point and prepared themselves for that. From the options available to them at the time, they made the best choices possible, in light of their current circumstances, the expectations of their discipline, preferred work environment and culture, and an understanding of their own values, needs, interests, strengths, and skills.

This approach reflects the contemporary view of career development professionals in the West that career management is "the lifelong process of

1

managing learning, work, leisure, and transitions in order to move toward a personally determined and evolving preferred future" (National Steering Committee for Career Development Guidelines and Standards, 2004, p. 139). Clearly, in this paradigm, the term *career* differs from a job or profession. More encompassing than either, it is a "lifestyle concept that involves the sequence of work, learning and leisure activities in which one engages throughout a lifetime [and in which one invests energy to create something that is bigger than oneself].[1] Careers are unique to each person and are dynamic; unfolding throughout life" (ibid).

The stories presented in this and following chapters illustrate just how individual each successful scientist's life is: how different the beginnings, the opportunities, the choices, the balance between professional and personal roles. We begin by examining the concept of success and identifying some common characteristics and attitudes of successful female scientists. The critical impact of our *context* on the pressures that we experience, constraints that we must deal with, and opportunities available to us is stressed, as is the need for each of us to seek out and, as much as possible, create opportunities for ourselves. The chapter closes with a discussion of strategies for managing our own careers.

WHAT IS SUCCESS?

We all need to feel that we are contributing to society in some positive way, that our efforts are valued and appreciated, that our work (whether paid or unpaid) *matters*. Though the expression of this need varies with the individual and her circumstances, it underlies all striving for success. But what, exactly, is *success*? The *Oxford English Dictionary* defines it as "(1) the accomplishment of an aim: a favourable outcome. (2) The attainment of wealth, fame or position. (3) A thing or person that turns out well" (Pearsall and Trumble, 1996, p. 1440). Clearly, it is important to distinguish between the criteria that *we* use to judge success[2] and those used by society,[3] for they may differ. How we can deal with a mismatch is discussed extensively throughout this book.[4] Of particular importance is a reaffirmation of our values, priorities, and goals[5] and the support of our network,[6] as the following story illustrates.

---

[1] This is how the terms *career* and *career management* are used throughout this book.

[2] See "Managing Your Career for Success" in this chapter.

[3] See "Realities of Context."

[4] Especially in the chapters "Climbing the Ladder" and "Balancing Professional and Personal Life."

[5] See the chapters "Time Stress" and "Mental Toughness."

[6] See the chapter "Networking."

During the final year of her PhD program in a top-ranked university in Canada, an engineering student began receiving tremendous pressure from her academic colleagues to apply for tenure-track positions in academe. In a discipline in which female faculty are underrepresented, they saw her as a "perfect fit": she is an excellent scientist, loves teaching, AND is female.

Though interested in an academic career, she knew that she wanted to gain industry experience first, so that when she entered academe, she could anchor her skills in real-life problems and be able to bring to her students perspective and experience from both the theoretical and applied worlds of engineering. At the same time (and on a more personal level), she was involved in a serious relationship with a partner who lived in a larger community several hours away with whom she was planning marriage and a family. To her mind, the best choice was to seek a position in industry, in the community where her future husband lived.

But she respected the experience and advice of her colleagues and superiors, was understandably flattered by their unreserved confidence in her, and—to a certain extent—enamored of the idea of becoming a faculty member—and an important role model—at such a young age. So she applied for several jobs and was offered positions even before finishing her Ph.D. This put her in a very awkward position because the interviews confirmed that her best choice after graduation would be to work in industry. When she respectfully declined, her superiors were surprised, some expressed disappointment and even anger; a few declared that she was ruining her career. But their feedback did not stop there. She was continually encouraged to reconsider her decision, and when she did not, she felt pressured to explain herself and justify her actions.

It was a difficult time for her, especially because she needed to focus all her energies and attention on completing her thesis and preparing for her defense. Under the strain, she began doubting not only her career decision, but even her abilities as a scientist. Fortunately, she had the support of trusted friends and family members to encourage her, and when she reexamined her own values and personal and professional goals, she was able to make the conscious choice to believe in her own reasoning and trust her decision.

Today she is newly married and happily employed in a stimulating position at a highly respected engineering consulting firm.

Successful Women Scientists: Shared Qualities, Common Themes

Working hard overcomes a whole lot of other obstacles. You can have unbelievable intelligence, you can have connections, you can have opportunities fall out of the sky. But in the end, hard work is the true, enduring characteristic of successful people.

Rear Admiral (Ret.) Marsha Evans, President and CEO, American Red Cross

Though each career is distinct, successful women in science share many qualities, attitudes, and goals that can be instructive to aspiring scientists. All have discovered that achieving in science—indeed, as in all of life—requires vision, focus, dedication, determination, a commitment to lifelong learning, and a persistent striving for excellence. It involves courage and a willingness to make difficult choices and compromises.

Engineer DR. LUCIANE CUNHA was working in industry in Brazil until an opportunity to do doctorate work lured her to the United States. She had written a national exam after completing her undergraduate degree and was one of only two women to be hired by Petrogas, a major petroleum-engineering firm in her country. Something of a pioneering woman in her field, she excelled in her work and was promoted to a position on an offshore oil rig (an environment that had no facilities for women), and she continued to achieve.

By the time she was offered a position in graduate school, she was established in her work, was married, and had a family. But the opportunity to pursue her research interests and advance her career was too good to decline. She accepted the position, even though it meant leaving her husband and children at home.

After earning her Ph.D., she was offered a permanent job in an academic institution with a strong, international reputation in her field, and moved to Canada—to a very different social structure and climate. This time, her children moved with her, while her husband remained in South America. Several years later, her husband secured an academic appointment at the same university and left Brazil to join the family.

For Luciane, the many hours of lost sleep, time she would like to have spent with her children, and missed opportunities for professional advancement in industry were necessary compromises. "Professionally speaking, I could have achieved a CEO position, had I stayed with Petrogas," she believes. "Sometimes I had to say 'no' to things like that because I was balancing professional and personal responsibilities." But she does not regret her choices. "I am happy with my career, especially in my roles as educator and mentor to the women in my classes. I try to teach them that their choices are not 'bad' ones, that they certainly will be able to accommodate all the things in life."

At times, achieving in science may mean disregarding well-meaning advice about what you cannot do and the way things have "always" been done. But most important, it involves remaining flexible.

DR. JUNE E. OSBORN, President of the Josiah Macy, Jr. Foundation, an American organization dedicated to improving the education of health professionals, also took advantage of unexpected options and opportunities, and ended up combining science and public policy. She chose medicine because she loved science, enjoyed people, and was good at interacting with them, and, more practically, because she saw medicine as a field that offered a wealth and breadth of job possibilities. Rather than open her own medical practice after she finished her pediatric training, she completed postdoctoral work in virology and pursued a career in academe at the University of Wisconsin.

At the time, the institution had a nepotism rule that prevented her from holding a primary appointment in the same department as her husband (also an academic), so she accepted a faculty position in the Department of Microbiology. After 18 years of research and teaching—and raising three children—it turned out that she had the ideal credentials to apply herself to the study of AIDS when it emerged. "It was a remarkably systematic coincidence. If someone had sat down two decades earlier and said 'we want to be ready to take on the world's greatest epidemic when it comes along,' they would have recommended the kind of educational preparation and research experience that I had. The saying 'chance favors the prepared mind' might apply to what happened to me."

As one of the few women in her field in those early years, she served on more than her share of professional groups and federal advisory committees. She was involved in the heated vaccine controversies of the 1970s and chaired the committee that advised the National Institutes of Health on the emerging AIDS epidemic in the early 1980s. These roles revealed her instincts for public policy, her ability to lead and not fold under pressure, and her intuitive ability to work with the media and translate complex scientific concepts into simple, comprehensible English. Her work gradually led her away from the laboratory and out of the classroom, to chairing the national commission on AIDS, serving as dean of a major school of public health, and more recently, serving as a foundation president. It has been, she observes, "a startlingly ecumenical career."

A similar theme echoes through the stories of other successful female scientists. Though each has an individual twist, all reveal how these women prepared for, recognized, seized, and even created their own opportunities, as illustrated in the next two stories.

As she reviews her professional life, DR. MARGARET-ANN ARMOUR, Associate Dean of Science (Diversity) at the University of Alberta (Canada) and coauthor of the chapter "Transitions", does not consider that she ever looked into the future and planned where she would be in a decade's time. Instead, she "responded to highly unexpected invitations" that arose from doing excellent work, being involved in the broader scientific community, and her ability to follow through on her belief that "I could make of a job what I wanted to make of it." When she started as a laboratory coordinator at the university some 25 years ago, she was one of three people on a safety committee. Together they came up with the idea of developing a set of procedures for safely handling waste materials that were simply thrown into the trash—there was no consideration for the impact on the environment at the time. This led to a commitment to find ways to transform hazardous into non-hazardous materials that was "not planned, but has been a huge part of our careers."

At an early age, DR. SHIRLEY ANN JACKSON, President of Rensselaer Polytechnic Institute (Troy, NY), demonstrated a scientific sensibility and natural curiosity that was encouraged by parents who believed that education was the basis for success and good citizenship. As she matured, her intellectual gifts, focus, and determination developed into a "can-do" approach to life that is unstoppable. Her experience as a female science student—and a woman of color—in a male-dominated discipline, at a predominantly male university, taught her to turn the potentially negative experience of being "different" into an opportunity to demonstrate her competence and be accepted for her excellence. She learned to guide her career by "pushing the envelope."

THE REALITIES OF CONTEXT: PRESSURES, CONTRAINTS, OPPORTUNITIES

Societal definitions of success vary considerably, depending on our context,[7] and with them, the expectations placed on us in our various roles in society. These will greatly affect the assessment of our performance, contributions, and even our worth. It is important to understand and accept this reality of life, for we will experience occasions when there is a mismatch between our values and those of society. More critically, perhaps, is the need to be wary of the potential negative effects of societal values on our own expectations and beliefs—about ourselves, what choices we have, and what we can achieve.

For example, in those low- and middle-income nations where women are expected to fulfill only the traditional roles of wife and mother, society's judgment of a woman's success is based solely on an assessment of her contributions to the support and nurturance of her family. The implications for women who aspire to a life in science are obvious. Even in Western societies, where traditional expectations of women are no longer as rigid, there remain vestiges of the belief that making a commitment to science is incompatible

---

[7] The "society" in which we live and work, be it geographic (e.g., country, region, community), organizational (e.g., business, institution, department or research group), or other (e.g., professional discipline; socioeconomic, racial, cultural, or age group).

with having a personal life that involves responsibilities for others (be they children, elderly parents, members of an extended family), or other personal interests and pursuits that demand significant time and attention.

Within science itself, the criteria for success vary with the environment in which one works (independent of gender). In research-intensive universities, for example, merit, promotion, and tenure are awarded according to a scientist's ability to attract research funding and graduate students, numbers of papers published, and even by the number of awards and honors. Contributions to teaching and service are given greater emphasis in those institutions that concentrate predominantly on training. In the commercial world, financial rewards and promotions are earned through the profitability of products and/or processes that result from research programs and the number of patents generated.

To manage our careers effectively, therefore, it is important for us to understand how "success" is defined and measured within our current context (or in new situations that we are considering) and to identify the underlying values of the system. The clearer our understanding, the easier it will be for us to identify and choose systems that are most compatible with our own values, priorities, and goals.

Unequal Opportunities

Our context also influences the number and range of opportunities available to us. For example, in those high-income nations where there are a strong and growing infrastructure and an increasing commitment of funding to research and development initiatives—by governments, industry, and philanthropic organizations—women scientists have many options for paid employment in academe, industry, business, government, and private research institutes. These cultures also tend to be sensitive to the need for, and benefits of, diversity, though access to opportunities is not yet universal. Encouragingly, many are engaged in efforts to actively recruit women and other minorities to positions at every level of scientific training and work, from undergraduate education to the highest research and management positions.

Other factors and trends are creating an even greater need for talented and committed scientists and are increasing the range of opportunities for women. The genomic and computer revolutions, as well as the new and emerging technological tools, offer great promise for scientific discovery and are transforming how we conduct research. Collaboration across scientific disciplines and national borders is expanding our avenues of inquiry and inspiring us to address the increasingly complex scientific and technological challenges confronting the world, such as global epidemics, antibiotic resistance in bacteria, the threat of bioterrorism, and environmental decline.

Prospects for careers in science will be even brighter in the future because of changing demographics and what some policymakers call "capacity issues": more senior scientists (in all sectors) are retiring than can be replaced because there are not enough qualified young scientists available.[8] For example, in the next few years at the National Aeronautics and Space Administration, 25% of the engineers will be eligible for retirement. The situation is similar in government, industry, and business outside North America.

Opportunities such as these are not available to all women across the globe. They are more limited for those living in societies where roles for men and women are more traditional, as mentioned earlier. Though views are changing (albeit slowly), and it is becoming more acceptable for woman to pursue higher education and even careers, other circumstances can severely restrict the possibilities for pursuing a career in science and/or hinder productivity.

For example, the population of a country and strength of its economy will affect the numbers and availability of jobs. Without jobs, female—as well as male—scientists may have no option but to leave their home countries to advance their careers. In Portugal, for example, most scientific research is conducted at universities. Faculty turnover is very low, and positions rarely become available. Opportunities in industry are almost nonexistent. This severely limits the options of postdoctoral fellows who must wait until someone leaves, retires, or dies before they can hope for any security in their chosen fields. Many early career scientists barely manage to support themselves and have to live on their own savings when they are between projects. These circumstances force a difficult choice: Many who want to stay in science leave Portugal; those who want to stay in Portugal leave science. Scientists in other countries face similar realities.

The lack of reliable infrastructure support presents a very different set of conditions and challenges. For example,

---

[8] Obviously, training more scientists is an important, long-term solution, but it takes years for individuals to develop the needed skills, knowledge, and experience. A complementary solution is to encourage women to stay in science and attract back to science those who have left. In their Foreword, Drs. Miriam Stewart and Mark Bisby discuss some of the data on the disappearance of women at successive career stages in health research in Canada and describe the efforts of senior representatives of the Canadian Institutes of Health Research and other key research funding agencies in Canada to foster success and sustainability in women's research careers. Governments in other high-income nations, such as the United States and the European Union, have introduced their own initiatives.

In the former Soviet Republic of Georgia, microbiologist DR. MZIA KUTATELADZE has been working at the Eliava Institute in Tbilisi (the home of the world's largest collection of bacteriophage against human bacterial diseases) since 1987. During the breakdown of the Soviet system, it was almost impossible to conduct research. Heat and electricity were unreliable, and no funding was available for research. Despite the hardships, she persisted. She applied for and was awarded two collaborative research grants from the North Atlantic Treaty Organization (NATO) that enabled her to work in the NATO laboratories at Toulouse (France) for a time; she has since returned to Georgia. She had to ask her international collaborators to lobby the American company supplying Georgia's electricity to put the institute at the top of the priority list, so that it would have a reliable power supply. Even now, as a senior scientist, there is little research support from the government; she depends almost entirely on grants from international organizations and what assistance she receives from her international collaborators.

Changing Contexts to Improve Opportunities

One of the choices women scientists are making to improve their opportunities for training and work is to leave their home countries, either temporarily or permanently.[9] Such decisions can be very difficult because they involve many compromises. Some, like the postdoctoral fellows from Nigeria and Uruguay who were interviewed for this project, knew that they had only one option if they wished to pursue their dreams of a life in science: to leave their homes and never return there to work. Both moved to Europe where there are better training opportunities in their respective fields and promising prospects for employment, but each felt—and were made to feel by some who remained in their home countries—that they were turning their backs on their families and rejecting their cultures. Another research scientist, Dr. Marianne Nyman, Assistant Professor in the Department of Civil and Environmental Engineering at Rensselaer Polytechnic Institute, moved her family from Europe to the United States because of the career opportunities for her and her husband. By doing so, she had to leave an advanced, free, child care system in her native Finland and the support of her extended family—a difficult trade-off indeed.

[9] This topic is discussed in detail in the chapter "Training and Working Abroad."

MANAGING YOUR CAREER FOR SUCCESS

If I have seen further it is by standing on the shoulders of giants.
Isaac Newton, in a letter to fellow scientist Robert Hooke (February 5, 1676)

Truly successful scientists build on the foundations of discovery laid down by those who have gone before them. Likewise in managing a career in science, the experiences and insights of mentors and examplars can be instructive to aspiring scientists. Though the stories in this and subsequent chapters can guide and inspire you, look also to your own mentors and the members of your own network as you face each new decision in the management of your career. Their advice, support, and encouragement will prove invaluable.

In addition, there are many excellent print[10] and online resources available to assist you in understanding the process and developing the necessary skills. One resource that many have found to be particularly helpful is the *Career Development eManual*, developed by the professionals at the Career Services Centre of the University of Waterloo, Canada (2005). You may also choose to consult knowledgeable specialists (e.g., guidance and career counselors, executive coaches) if you wish to take advantage of group or individualized training sessions. Dr. Christine Faerber discusses one such training program in the chapter that follows and describes some of the key benefits of participation, expressed in the participants' own words. Other contributors, including her coauthor, Dr. Anngienetta Johnson, have benefited from the advice of "career coaches." All would agree that being deliberate about managing your own career is an important factor in achieving success. Though each may describe the process slightly differently, generally speaking, career management involves four steps: (1) self-awareness, (2) opportunity awareness, (3) decision making and planning, and (4) implementation and periodic review.

Self-Awareness

The first step in effective career management is self-assessment. When we are clear about what success means to us (based on our values, needs, and preferences) and have an understanding of the professional and personal resources that we have to offer, we will be better able to identify, evaluate, and create opportunities that are the best match for us.

Many checklists and inventories exist to help you identify your values, needs, personality preferences, attitudes, strengths and weaknesses, learn-

---

[10] For example, Rosen, S. and Paul, C. *Career Renewal. Tools for Scientists and Technical Professionals*, New York: Academic Press, 1997.

ing and conflict resolution styles, and the like. These tools are not intended to be prescriptive, as Margaret Riley Dikel points out in her CareerJournal.com article "A Guide to Going Online for Self-Assessment Tools" (2005), but to enhance self-knowledge and provide criteria on which to base career decisions. With such awareness, you will be able to make the best choices possible at the time and follow-through with confidence.

What Is Important To Me? What Are My Needs?

Most of us already know the answers to the questions "What is important to me?" "What do I value? What do I believe in?" "What do I need for a fulfilling life?" for they are central to who we are. It is just that we do not always spend time reflecting on them. But our core values, beliefs, and principles—the foundation of "core mental strength" (defined in the chapter "Mental Toughness")—and our core needs are the basis for determining our priorities and setting professional and personal goals. If you do not take your values and needs into account when making job choices, you may choose unwisely and end up disliking your work—a consequence that will compromise your ability to succeed.

Values clarification exercises, such as the one presented in the *Career Development eManual*, can also be useful in helping us to articulate and examine more closely what is important in our lives and what will, ultimately, give our lives meaning. Some scientists develop a "mission statement" for their lives that informs all the decisions that they make. For Dr. Jeanette Holden, Director of the Cytogenetics and DNA Research Laboratory at Ongwanada (Ontario, Canada), it was her personal experience of her brother living with autism that inspired her to commit herself to the study of genetic disorders associated with developmental disabilities and her special interest in autism. Others, such as Dr. Kathleen Sendall, Senior Vice President, North American Gas, Petro-Canada, have identified key inspirational quotes that express their core values and principles (Sendall, 2000). Many refer to these written statements on a regular basis to help them remain focused on what is important to them.

What Truly Captures My Interest? What Do I Enjoy Doing? What Excites Me about Science?

Find what you enjoy doing and follow that path, for it will take you to a place where you will be happy.

Rusty Schweiker, former American astronaut

Our interests can be reliable indicators of what is important to us and offer clues to what naturally engages us and inspires us to action. When we choose to invest our time and energy in activities that are related to our interests, we have a natural stamina that will see us through the inevitable frustrations and challenges of research and the political aspects of science. Identifying your interests is simple: Observe what you naturally choose to do when you are free of pressing commitments and feeling most "yourself." Ask yourself questions such as "What do I enjoy doing?" "How do I prefer to spend my free time (when I have it)?" "What am I drawn to without even thinking about it?" "What excites me about science?" "What kinds of books, magazines, journals, and newspapers do I read?" "What are my favorite television programs? Movies?" "What do I enjoy thinking about? Discussing?" The answers to these and similar questions can provide insight into the kinds of jobs, work environments, and training activities that you may prefer. When we pursue what we love to do, we will become good at it. We will develop greater confidence in our abilities and choices, and ultimately, achieve success.

What Have I to Contribute?

We must believe that we are gifted for something, and that this thing, at whatever cost, must be attained.
     Marie Curie, winner of Nobel prizes in Physics and Chemistry

When we identify for ourselves the skills, knowledge, and experience that we currently have to offer, we will be able to articulate more effectively how we can contribute to society through our professional and personal work. This is what potential employers are most interested in when evaluating applications and interviewing candidates. Not only are the resources that we bring from our formal training in our chosen fields important to success in our chosen fields, but also the qualities of mental toughness, personal management, and the other complementary skills and knowledge that are discussed in this book.

Though we cannot accurately predict what competencies will be needed in the next decade, we may be able to anticipate what we will need to take the next step in our unfolding careers. By beginning *now* to acquire the new skills and knowledge in the areas of our developing interest, we will be more prepared to pursue unexpected opportunities as they arise. This is an excellent strategy for achieving the breadth and flexibility that we will need to be successful.

What Does "Success" Mean to Me?

Success consists of the progressive realization of predetermined, worthwhile goals.

Paul J. Meyer, American philanthropist and motivational speaker

As important as it is to understand how "success" is defined and evaluated within our current context, it is equally important—if not more so—to define "success" for ourselves. Our judgment of how we meet our own expectations can have a profound impact on our motivation, confidence, and self-esteem.

Many people think of success in terms of goals. But, as Joanne Lozar Glenn describes in *Mentor Me: A Guide to Being Your Own Best Advocate in the Workplace* (2003), these goals need to be meaningful to you. Ask yourself questions such as "What is the legacy that I want to leave and how will what I am doing contribute to that legacy? What makes me happy? What must I do to be fulfilled?" Your goals may be part of a lifetime mission or passion (as Nadia Rosenthal so eloquently expressed in her Prologue) or may change with age and experience, as the following story illustrates.

DR. JOANNE SIMPSON's goal was not that precise when, at the age of 14, she said to herself "I am going to get somewhere and be somebody." She had no idea at the time of where or what. She simply was determined to succeed. She did, eventually transforming her early fascination with clouds into a career as a world-renowned meteorologist who, now in her 80s, still serves as senior scientist at NASA's Goddard Space Flight Center in Greenbelt, MD.

Her success came from hard work and determination and the fact that she trained herself to take advantage of opportunities as they arose. She studied history and mathematics in college but with no particular goal in mind. She was fascinated by aviation but lacked piloting skills or the eyesight to become a commercial pilot. With the outbreak of World War II, she joined the U.S. Navy WAVES (Women Accepted for Volunteer Emergency Service) where she learned about meteorology (under the tutelage of Swedish meteorologist Carl Rossby), and trained aviation cadets to forecast weather.

After the war, she wanted to study meteorology, but Rossby warned her "no woman has ever obtained a Ph.D. in meteorology. None ever will." She tried a few classes in more traditional fields for women, such as sociology and psychology, found them boring, and went back to meteorology. She eventually was accepted as a doctoral candidate at the University of Chicago and began the career that led to "so many honors and awards that my walls are covered with them."

In hindsight, Dr. Simpson concedes that her motivation for hard work in the early years of her career was inspired more by fear than by dreams of success: "If I failed, it would make it more difficult for the younger women to find opportunities." By the 1990s, though, so many opportunities in science had become available to women that she felt she could retire as a role model.

Opportunity Awareness

As I reflect on my life in science, I recognize that being prepared—and being willing to try the unexpected—were pivotal.
Dr. Shirley Ann Jackson, President, Rensselaer Polytechnic Institute

Through our own research efforts and by consulting our mentors and the members of our network for information, advice, and referrals (see the chapters "Mentoring" and "Networking" for strategies), we will be able to identify our options (traditional and nontraditional) in our current context and beyond. Our investigations may involve an examination of career trends, occupational information, industry and labor market information, new work alternatives, opportunities for continuing professional development, sources of funding for training and development, and the like. We also need to examine the implications of each option for our professional and personal life (e.g., level of intellectual challenge, opportunities for advancing science, remuneration and benefits, opportunities for partner, lifestyle) and how to address them.

Decision Making and Planning

Trust yourself. Create the kind of self that you will be happy to live with all your life. Make the most of yourself by fanning the tiny, inner sparks of possibility into flames of achievement.
Golda Meir, first woman Prime Minister of Israel

With a thorough knowledge of the options available to us, we can make informed choices based on our assessment of what will be the best match for our goals, preferences, and interests. There is no "perfect" choice. In the end, we will have to make our decisions with as full an understanding of the implications and consequences as possible and to accept the inevitable compromises that will have to be made in the imperfect systems in which we work.

American astronaut Dr. JANICE VOSS decided she wanted to go into space when she was in the fifth grade after reading Madeleine L'Engle's childhood classic *A Wrinkle in Time* (1962). It was not until she was in high school that she realized that this meant becoming an astronaut. But it became and remained her fixed goal, one that she achieved in 1990. She admits that her path, which appears to be a straight line from aspiration to fulfillment, always involved having an alternate plan that she occasionally followed. Her academic work, from her master's degree through her doctorate, was strained by a series of canceled projects and programs, departed and deceased advisors, all of which she surmounted to emerge with a Ph.D. in aeronautics/astronautics from the Massachusetts Institute of Technology. She decided not to apply for the astronaut program then but turned instead to "Plan B" and accepted a job with the newly formed Orbital Sciences Corporation. This decision turned out unexpectedly well, for no new applicants to the space program were reviewed that year. When she finally did apply in 1990, she was accepted. As a member of the astronaut corps, she has flown in space, now, five times.

Once committed, we need to set goals accordingly, and plan our next steps,[11] as Dr. Voss did. These may include pursuing educational opportunities to acquire the necessary skills, knowledge, and/or experience that will strengthen our portfolios,[12] letting go of some responsibilities to assume new ones,[13] or conducting a job search.[14] Inevitably, the decision-making process has implications for balancing our professional and personal lives.

Implementation and Periodic Review

Winning the [Nobel] prize wasn't half as fun as doing the work itself.
Maria Goeppert Mayer, Nobel Prize–winning physicist

The final step in the career management process is following through on your decision, implementing your plan for achieving your goals, and periodically assessing whether your current situation still is the best "fit." The external changes that result from making a new choice are often accompanied by a period of inner "transition," when we come to terms with the new situation. As the chapter "Transitions" describes: "Unless transition occurs, change will

[11] The chapter "Time Stress" offers strategies for goal setting that will assist you in this process.
[12] The next chapter ("Continuing Professional Development") discusses this topic in detail.
[13] See the chapter "Climbing the Ladder."
[14] For some general strategies, refer to the *Career Development eManual.*

not work" (Bridges, 1991, pp. 3–4). It is during this period that you need to use the strategies of mental toughness to let go of doubt and insecurity, stay focused on your goals, and remain patient and confident in your decision.

> *I must admit that I personally measure success in terms of the contributions an individual makes to her or his fellow human beings.*
>
> Margaret Mead, anthropologist

The process of assessing how satisfied you are with your current job and evaluating how well it matches your personal and professional qualities and goals is a potentially stressful—though necessary—part of the process. It is during your review (which may occur every few years) that you will be able to readjust some of your goals (if they are no longer relevant), add new ones (if some have been achieved), or identify when you need to make a significant change to your circumstances. This is the dynamic aspect of career management that will enable you to continue to develop as a scientist, remain responsive to emerging opportunities, and succeed.

In order to affirm your work in the context of your life—the "big picture"—you must invest in yourself. Be introspective. Ask yourself some tough questions. "What are your values? What do you stand for? What are the values of your organization? Are your values and those of your organization in alignment? Why do you choose to work in your current job and organization? Is there a fit? What are your talents and skills? What do you love to do?" My coach asks the question: "Where is the juice? What really 'jazzes' you? Are you doing it? What are your strengths? What things do you need to work on?" Once you examine first your values and their integration with your work, explore your strengths and weaknesses, and get clarity about what you love to do, you will make a great deal of progress toward discovering your life's work.

I personally did not do this self-exploration until after I had my first child. It is never too late, but certainly it is never too early. I encourage you to examine your values in the context of your career choices. Set yourself on a path of powerful learning. Your opportunities for growth and development are tremendous. Most significant to your success will be your willingness to invest in yourself. Seek the insights and help of others in your journey. Remember, life is indeed the journey, not the destination.

KRISTI BROWN, Special Assistant for Strategy and Development, Goddard Space Flight Center

REFERENCES

Bridges, W. *Managing Transitions. Making the Most of Change*. Reading, MA: Addison-Wesley, 1991.

L'Engle, M. *A Wrinkle in Time*. New York: Dell, 1962.

Lozar Glenn, J. *Mentor Me: A Guide to Being Your Own Best Advocate in the Workplace*. Reston, VA: National Business Education Association, 2003.

National Steering Committee for Career Development Guidelines and Standards. "Glossary of Career Development Terms," *Canadian Standards and Guidelines for Career Development Practitioners* (2004) [Online]. Available at: http://www.career-dev-guidelines.org (accessed August 20, 2005).

Pearsall, J., Trumble, B. (eds.). *Oxford English Reference Dictionary*, 2nd ed. New York: Oxford University Press, 1996, p. 1440.

Riley Dikel, M. "A Guide to Going Online For Self-Assessment Tools," *CareerJournal.com. The Wall Street Journal Executive Career Site* (2005) [Online]. Available at: http://www.careerjournal.com/jobhunting/usingnet/20030429-dikel.html, 2005 (accessed August 20, 2005).

Rosen, S., and Paul, C. *Career Renewal. Tools for Scientists and Technical Professionals*. New York: Academic Press, 1997.

Sendall, K. *New Frontiers, New Traditions, Kathy Sendall Keynote Address. July 6, 2000* [Online]. Available at: http://www.mun.ca/cwse/Sendall.pdf (accessed August 20, 2005).

University of Waterloo Career Services. *Career Development eManual* (2005) [Online]. Available at: http://www.cdm.uwaterloo.ca (accessed August 20, 2005).

●●●●●●●●●●●●●●●●●●●●●●●●●●●●●●●●●●●●●●●

CHRISTINE FAERBER, Ph.D.
Head, Competence Consulting, GERMANY
Political Scientist

●●●●●●●●●●●●●●●●●●●●●●●●●●●●●●●●●●●●

Dr. Christine Faerber is one of Europe's leading experts on women in science and gender politics. She started her professional career at the Free University of Berlin as head of the equal opportunities office. During her 8 years in this position, Dr. Faerber was elected President of the State and Federal Organizations of Women's Representatives at German Universities. She held several offices in other academic organizations, the most important of which were her membership in the Commission for University Development in Berlin and in a commission of the German Conference of University Presidents. Her experience includes a broad view on transformation processes for women in science in the former communist states, especially in Eastern Germany, and on the participation of women in the research programs of the European Union.

In 1999, Dr. Faerber founded her own research and consulting institute to be able to help (and move!) ministries and universities to introduce gender aspects into their administrative and scientific tasks. The second and equally important focus of her work is training, mentoring, and consulting for women in science. Dr. Faerber has worked with more than 1000 women scientists, most of them from the German-speaking parts of Europe.

Being from a rural part of southern Germany, the Black Forest, Dr. Faerber has experienced the different roles for women and opinions on public child care in Northern, Southern, Eastern, and Western Europe. Most women in her country, even well-trained professionals, stay at home with their children or work only part time. The good day care conditions and the high acceptance of women professionals around Berlin make it possible for Dr. Faerber to be able to enjoy her family with two children and lead her firm with its challenging mission to combine excellent research and highly relevant political practice.

ANNGIENETTA JOHNSON, D.Sc.
Assistant Associate Administrator for Education
National Aeronautics and Space Administration, UNITED STATES
Expert in Engineering Management & Systems Engineering

Dr. Anngienetta Johnson is the consummate lifelong learner. Her illustrious career began at Texas Woman's University where she earned a BA in mathematics (1971). As a cooperative education student, she complemented her academic studies with a work tour at the National Aeronautics and Space Administration (NASA). Since then, she has continued to develop new skills, knowledge, and expertise through informal and formal learning opportunities at NASA and beyond, including earning two master's degrees: in Industrial Management (University of Houston) and in Information Systems Management (George Washington University). Her determination to help those in crisis led her to pursue a doctorate of science degree from George Washington University's Institute for Crisis, Disaster, and Risk Management. This approach to professional development—combining formal learning with practical experience—is one of the hallmarks of her success.

During her 36 years at NASA, Dr. Johnson has held a variety of positions, including managing the development of Earth orbiting spacecraft and overseeing NASA's institutional and informational assets. Since 1994, she twice received Peer Excellence and Spaceship Earth Awards and is the recipient of the 1998 NASA Headquarters Creative Management Award. In a recent move, Dr. Johnson accepted the challenge to be the advocate and spokesperson for research that advances promising science, technology, engineering, and mathematic (STEM) educational concepts and practices. These efforts will help provide today's students with a path toward rewarding careers in STEM and will create a workforce capable of exploring the moon, Mars, and beyond.

Serving people is something Dr. Johnson does regularly and with passion. Her tireless efforts to inspire women to achieve professionally in science and technology have not gone unnoticed. She was named Outstanding Woman of the Decade in 1977. That devotion to women's advancement is equaled by her commitment to community work. It is quite common to find her comforting victims of fire, floods, tornados, or hurricanes. She is also an enthusiastic advocate for the homeless, people living with disabilities, and the addicted. In 2004, the President's Council on Service and Civic Participation recognized her significant contributions with the Volunteer Service Award.

Chapter

2

· · · · · · · · · ·

CONTINUING PROFESSIONAL
DEVELOPMENT

· · · · · · · · · · · · · · · ·

Christine Faerber and Anngienetta Johnson

> *Education multiplies one's options and opportunities in life. Why limit your possibilities, when there is an exciting world out there waiting for your brainpower?*
> Dr. Shirley Ann Jackson, President, Rensselaer Polytechnic Institute

Your career will likely be a journey with many twists and turns, as you respond to advances in your field, make new choices that are consistent with your changing priorities and goals, and pursue emerging opportunities for professional and personal development. Learning does not end when you complete your doctorate or postdoctoral fellowship. It continues throughout your life. By developing the attitude of a lifelong learner and acquiring new skills and knowledge as your responsibilities and interests dictate, you will be better equipped to recognize, seize, and even create opportunities that will lead to greater fulfillment and success. It is never too late to expand your horizons. You have begun already: by reading this book, you are being proactive in the management of your career and involving yourself in ongoing learning.

Continuing professional development involves (1) staying current, skilled, and intellectually fresh in one's chosen discipline (i.e., developing scientific skills and knowledge) and (2) developing complementary knowledge and skills, often called "employability skills" in North America. These are "the skills, attitudes and behaviours that you need to participate and progress in today's dynamic world of work" (Conference Board of Canada, 2005). They are not job specific but cut across all types of work. They will enhance your ability to function effectively as an individual scientist and as a member of the community of science, and include fundamental skills (i.e., communication, information management, numeric literacy, critical thinking, problem solving), personal management skills (i.e., positive attitudes and behaviors [including stress management], responsibility, adaptability), and interpersonal competence (e.g., teamwork).

Anticipating the skills and knowledge we will need at each new next stage in our careers requires self-awareness, knowledge of developments and trends (both discussed in the chapter "Career Management"), imagination, and the self-confidence that comes from previous successful experiences. Obviously, when we are starting out we have less experience, but this will grow with time. We are not alone in the process. Our Mentors, peers, and members of our network are invaluable sources of information, guidance, and support, as you will read in the next two sections. Christine Faerber begins by stressing the importance and benefits of training programs designed specifically for women through the personal stories and comments of participants of a four-year career development initiative sponsored by the German government.

There are many factors influencing the number, type, and range of learning opportunities that are available to us, just as there are for work opportunities. The *context* in which we are living and working is an important one that was discussed in detail in the previous chapter. This includes the institutional culture (e.g., value of and rewards for continuing professional development) as well as the degree of flexibility allowed by our jobs to participate in educational initiatives. The availability of time and funding are two other crucial factors. In the section on strategies you will read how female scientists and engineers took advantage of the opportunities at one organization, the National Aeronautics and Space Administrations (NASA), to create very different, yet successful careers. The discussion continues in the next chapter, "Training and Working Abroad."

THE BENEFITS OF TRAINING PROGRAMS

Women are very skilled in the scientific and technical aspects of professional work, and very naïve about the other aspects of scientific life. The development of complementary skills is critical.

The experience of European countries like Germany and Switzerland has clearly demonstrated that specific training and mentoring programs for women in research and higher education can contribute greatly to their achievement and success. One such program, initiated in 2000 by the German Minister of Education and Science and entitled "Impulse for Advancement," provided training for more than 800 participants (Bundesministerium für Bildung und Forschung, 2001, 2003; Center of Excellence Women in Science Program, 2004). All were highly qualified career scientists who had completed their postdoctoral studies, most had even published a second book or completed the "habilitation,"[1] and some participants also held permanent positions in academe. The program was conducted from 2001 to 2004 and provided three-day training sessions involving reflection, strategy building, and networking for groups of 12 to 20 women and included individual follow-up coaching sessions.

More than 1,000 women scientists from the 16 states within the federal republic of Germany applied for this program. Many of them had no previous experience with training programs conducted exclusively for women, which explains why many participants were uncertain about the benefits.

> *To be honest, I thought, "What would be the use of a women's training program?" when I finally received the invitation. In my field men decide who becomes a professor.*
>
> Engineer, 42

This remark, written on the final feedback form, was typical, but the participants' comments continued:

> *Never in my life have I interacted with other women scientists in such a way or under such circumstances. I had always thought that my problems were unique. I have learned that there are structures that I can learn to handle, that there are competent women who can and want to support me. I have learned that I can prepare myself and manage my career better than I used to do. I am very grateful that I was able to participate in this course.*

Clearly, the experience had a positive effect on this participant. From the feedback and the follow-up interviews of many others, we know that this effect is not singular.

The Minister had set herself an ambitious aim when she came into office in 1998: She wanted to increase the percentage of women in professorships in Germany, from less than 10% to more than 20%, within an

---

[1] A habilitation is an academic examination following the dissertation. It requires the publication of a second book or published research with a high impact factor.

eight-year period. This goal was overambitious (Faerber, 2002). The number of women professors has not increased to more than 12%, and Germany is still at the very last position in the European Union when it comes to female participation in the permanent and well-paid ranks of academe. On the other hand, the goal of the specific supportive training program, i.e., to make women aware that they need to (1) focus on developing their careers (and not only on the purely scientific aspects of their work), (2) gather information about the systems that they work in or want to apply for, and (3) take preparative actions to support their professional advancement, was most successful.

The research on women in higher education in Europe has been advanced by a group of experts who examined the representation of women in the different member and candidate states in the union (Rees, 2002). The statistical comparisons show that the representation of women is extremely low in Germany, Austria, and The Netherlands and highest in Finland and Portugal. The numbers suggest that academe is a highly diverse and an extremely competitive field in Europe. Making a career in research and higher education is becoming more difficult with the expansion of higher education. Long hours and high-quality work are not the only factors that lead to a successful academic career; good connections and effective communication seem to be as important as the core academic performance. Yet many participants of the German training program very much underestimated the importance of these factors to their success. The large majority focused on the content and results of their research and did not invest their energies in developing complementary skills such as networking, self-promotion and presentation.

> *I thought the best aspect [of the course] were the videotaped sessions. I could actually see how much better I could have performed in the past if I had prepared better for the interviews (for full professorships, etc.). All these issues on self-presentation, understanding the expectations of the interviewing panels, etc. were things I had not considered to be so important before.*
>
>   Neurologist, 40, Senior Assistant

Many women academics feel embarrassed by the prospect of marketing themselves.

> *I thought "Why care about all this wrapping; I am no Christmas gift. People can see who I am and what I am capable of from my research publications." But now I know that I was not effectively communicating my achievements. I made it so easy for others to ignore me, to put me aside.*
>
>   Physicist, 37

More effective communication and networking are of vital importance to increasing the representation of women in professorships. All over Europe, and particularly in the countries where academic positions are highly esteemed, well paid, and academic freedom is high, there is an underrepresentation of women in higher positions, in spite of there being many excellent, well-trained, female researchers and academic teachers in more junior positions. This loss and neglect of the female potential are, to a large extent, due to the structures of academe that have been designed around male privileges and male role models.

One example is the negation of private life. The overwhelming majority of female scientists in Germany do not have children. More than 80% of the partners of these women are scientists themselves, and for the women, it amounts to a choice of career versus motherhood. While at the same time, most male professors "naturally" have children, and many have wives who do not work outside the home or who work part time or in highly flexible jobs, so that child care and household are still their responsibility.

> *I have never seen other women scientists with children. The experience of the other women in this course encourages me to dare starting a family myself. I have always wanted children and an academic career.*
> Life Scientist, 32, after her postdoctoral phase

This woman participated in a course that had among the participants four single women, seven women who already were mothers, and two high-achieving academic scientists who were pregnant—an unusual assembly in the German context. Her comments reveal another important function of the training programs that goes much deeper than the acquisition of book knowledge: When women actually spend time—playful, serious, intense time—with other women who are at the same career level, and, if the training is conducted well and the proper atmosphere created, mutual empowerment occurs.

Empowerment is crucial because the system and the individuals' experiences in the system are often so very frustrating for brilliant women academics. Not only do the structures negate motherhood, private life, family ties, and friendship, they also do not view women as successful professors, as scientific geniuses. Excellence is not attributed to women; diligence, hard work, persistence, and teaching qualities are, but brilliance is not (Brouns, 2003). In the training programs, the women interacted with other brilliant women and heard about the difficulties they experienced being accepted in their academic field. They recognized that the problem is not with the quality or quantity of their colleagues' work, but with a system in which a woman in the higher scientific ranks is easily disregarded by her male peers. Through these interactions, the participants of the workshops recognized the importance of networking and self-presentation for others, and, thus, could more easily accept the necessity of self-marketing for themselves.

Such a discussion cannot be set on the agenda too openly. For many participants, topics like "self reflection," "empowerment," and "role games" sound too esoteric, especially to academics who are accustomed to working with their heads and logic and not so much with their bodies and souls.

> *When I received the invitation I thought I should cancel the seminar. I have a lot of experience in academia and had already reflected much on my situation. So I expected nothing. But now I have so many interesting aspects to think about, from the trainers and from the entire group of scientists. I am highly satisfied that I attended. Actually I should have done something like this 10 years ago. Such a program should have existed when I was younger.*
>
> Pedagogue, 43, Habilitation in 1998

How do these training sessions enable women to have open and direct exchanges and develop career management strategies? What do the follow-up sessions contribute?

The agenda of the training session of the German program includes three aspects: First, it covers discussions and experience of selection processes in academe; second, it includes an exchange with senior scientists; and third, it offers individual counseling and feedback.

The first day begins with introductions on the backgrounds and expectations of the participants. This is followed by group work on the structures of selection processes in academe. Here knowledge on the diverse structures within Germany and in other countries is provided. The importance of networking and strategic performance is stressed by the trainers and supported with research data and specific examples.

> *I never knew how to contact people and whom to contact during an application process in Germany. My mentor has informed me I should not telephone. But he did not phone for me either. Now I know he has made it so difficult for me to apply elsewhere.*
>
> Senior Research Assistant, 35, Habilitation at the age of 27(!)

The afternoon of the first day is spent on the written job application, its structure, and the "do's and don'ts," with initial (general) feedback and group discussions and then individual, confidential feedback on the application documents of each. The evening is filled with stories of the participants as they reflect on moments of excellence. It is not surprising, but always a revelation, that most women academics (about 85% of the participants) chose as their best moment in their career a discussion after a presentation that they gave at a conference.

This sets the tone for the activities of the following day, which focus on the applicant's lecture for a professorship. The session starts with group work on the expectations that the university has of candidates for a new academic position—or rather, that the individual members of the university (e.g., selection committee, potential colleagues) have.

> *I had never thought about the interests on the other side of the table. Now that we have done this, it is all so clear. When I applied to a university in a small community, I did not take into account the interests of the people who would be my colleagues. I come from Berlin, and now I realize that they must have thought I'd apply, only to leave a few years later.*
>
> Natural Scientist, 37, Habilitation

The message of the session is clear: Be aware of the context of the lecture you give as part of your job application. This is not a conference talk, nor is it an ordinary lecture at your home university.

All participants present sample lectures to the group. These are videotaped so that they could review their performances, identify weaknesses, discover possibilities for improvement, and learn new approaches from the examples of their workshop colleagues.

> *So many of the women in this course were already so very competent. Now I see that there is always room for improvement. We have learned so many very concrete strategies on how we can improve our presentation skills. Better still, each of us has learned how she can improve her performance because the feedback was so individually suited and we are all so different. Now I have the information and insights I need to optimize my own performance!*
>
> Biochemistry Senior Assistant, 39

In the afternoon of the second day, the members of the teaching team meet individually with the participants to discuss their portfolio and the possible focus of future, one-on-one coaching sessions. (The program includes two follow-up sessions of 45 minutes each.) These can be used to discuss specific application processes, optimize the application documents, train for an interview, or discuss other relevant issues defined by the participant.

The evening of this intense second day is spent with a senior woman professor, sometimes with more than one guest. The participants have opportunities to ask questions on topics of interest and concern, including career planning, networking, the "inside" of academe, and balancing professional life and personal life. These professors have been carefully selected: they are women who will answer the participants' questions honestly and openly and who are supportive of young women academics (not all senior academics are).

One important issue for career planning in academe that always arises is the question of whether one should focus on science exclusively.

> *I thought it was so helpful that the Professor had said she always had Plan B—that is, she planned to become a professor, but always had another scenario in mind, another realistic option. I think in my field, this is important. There are so many brilliant people and so few positions.*
>
> Historian, 35, Postdoctoral Fellow

Another issue typically raised in these sessions is the question about partnership and children. Often the discussions revealed that the older generation had no role models, but that solidarity with other women (which has a lot to do with understanding each other's private situation) was helpful. In the following example, the workshop participant gained more insight into the difficulties that she was having with an existing mentoring relationship.

> *I was discouraged when I heard that the Professor and her partner live 600 km apart. This is not a lifestyle for me. But now I can understand my own mentor better. I think she has a very tough personal situation, and she must be envious that I have a husband and children and want to become a professor, too. She seems to think you have to give up the other aspects of life when you become an academic.*
>
> English Literature Senior Assistant, 38, Habilitation

The third and final day of the course focuses on mock interviews. In Germany, Switzerland, and Austria, interviews usually follow the applicants' lectures and focus on organizational questions. This is a different approach than in other countries (where candidates go through several interviews with larger groups or individuals for two or more days), so all the pressure is on this one interview. Therefore, much depends on the candidate's performance and ability to communicate in such an artificial situation.

Not only do the workshop participants take on the role of job candidate, but they also play the part of the members of the interview panel, and thus they develop a feeling for what it is like to be on the other side. The group work is intense and very constructive. Like the lecture presentation sessions, the interviews are videotaped. Participants clearly see that much of the success in the interview process depends on the applicant's ability to communicate effectively, and they realize that these strategies can be learned and developed.

> *Now, after analyzing the videotapes, I can understand why I did not get any further with my previous application. The colleague who played*

the applicant in the training session used the same strategy I had in real life. I did not tell the interview panel enough, and if they had not read my documents fully, they could not know about all my achievements. I had not told them about my five years abroad, my leadership position, about my most important publication, and my most expensive project...

Architect, 42

The three-day session concluded with oral and standardized written feedback from the participants. Most found the program to be highly relevant to their careers. The experience of self-reflection, group and expert support in the exercises, and video analyses were regarded as very helpful for the participants' self-management and personal development.

I have learned that I can prepare myself, that I am not just a victim, that I don't enter a black box. It seems an enormous amount of work and I think I have to sort it out a little, but I have the knowledge that I can contribute to my success. This is very empowering indeed.

German Literature Specialist, 37, Habilitation

The post-workshop follow-up sessions offered the participants opportunities to receive individual counseling from experts on career planning and the academic job search.

Workshop participants were not the only people to benefit from the sessions. Their experiences were recorded by the instructors and added to the growing collection of women's personal stories and experiences that enhance the trainers' work with other women who want to be successful in their academic careers.

The most important benefits of the "Impulse for Advancement" training programs were threefold. (1) The highly qualified women academics learned to value one another, to support others and receive and accept support, and to network, all are vital aspects of solidarity that are required if new women professors are to learn from other women and achieve success. (2) The participants learned the value of consulting independent career development and job search professionals. (3) Of course, the most immediate and tangible benefits were the improvements in their own performance in presentations and interviews.

When we are filling new faculty positions, I know exactly who has taken part in your training program. The performance of these women is SO good.

Equal Opportunities Officer at a university in northeast Germany, 2002

CONTINUING PROFESSIONAL DEVELOPMENT STRATEGIES

We must become the change we want to see in the world.

Mahatma Gandhi

The most sensible strategy for staying current and skilled is to choose (as much as is possible) an area that will be relevant in the future, learn which skills will be needed and which will become superfluous, and continually seek appropriate training to prepare yourself for each new step. Many women scientists have successfully used this strategy in their careers and gained useful insights for those who would come after them. Though their specific choices of learning opportunities differ, the results are the same. Most agree that continuing professional development is not enough; they also attribute some of their success to the advice and support they received from other women within the system and beyond (i.e., through their mentors and members of their networks).

The possibilities for developing strong scientific skills and knowledge, as well as complementary skills, are many and varied. They include independent reading of the literature (in your field and related areas); attending scientific conferences; participating in tutorials, workshops, short courses (such as the one described in the previous section), and summer institutes; enrolling in specialized degree and diploma programs; organizing visits to other laboratories to learn new techniques; and taking advantage of sabbatical leaves. These opportunities may be offered by employers, professional associations, manufacturers of scientific equipment, businesses, governmental agencies, independent training organizations, or educational institutes. Many women scientists at NASA have successfully taken advantage of these opportunities to advance their own careers. Their experiences can serve as examples. Here, they tell their own stories.

Developing Strong Scientific Skills and Knowledge

NASA understands what it takes to be successful and provides opportunities for employees who want to excel. One of its primary missions is to "inspire the next generation of explorers as only NASA can." NASA begins the process early and continues to inspire, train, and influence its personnel throughout their careers. They have access to a cadre of NASA-provided (as well as external) training and self-development programs. As a consequence, the organization continues to produce some of the most successful women in scientific and engineering fields.

I entered a graduate program in materials science immediately after completing my undergraduate degree in chemistry. I loved this area of study, but had not thought about how I might apply my new knowledge toward a meaningful career after I obtained my master's degree. One afternoon I mentioned this to my research advisor. I also told him that I was a bit apprehensive about being female entering in a largely male-dominated field. It was several days later that he handed me some documents to read. They contained information on NASA's summer graduate student program that sparked my interest. I applied and was accepted. I can definitely say the experience changed my life. I worked at Langley Research Center for 16 weeks and experienced what it was like to work in a research laboratory. I actually applied my skills in a "real-world" environment and realized my strengths and weaknesses. When I returned to school the following autumn, I was able to focus on strengthening and balancing my skills. The NASA graduate student summer program gave me insight into the world of work that I would never have had if I had limited myself exclusively to a traditional classroom setting. It was the application of what I was learning that made all the difference for me. My research advisor was my mentor, something that is so important to maximizing your academic development and preparing adequately for your future.

GALE ALLEN, Ph.D., Exploration Systems Mission Development.

I began my career as a physics major. Though the course work was challenging, I was fortunate to have professors who made it interesting. Since I was putting myself through school, I had a vested interest in achieving success and thus put a lot of effort into my studies. One of my part-time jobs throughout the school year was in the seismology laboratory. Not only was I putting to use the skills I was learning in the classroom, I was applying them in different areas. Through this experience, I was able to decide which area of specialty interested me most.

After graduation, I decided that I still had much to learn before applying for full-time work, so I attended graduate school and earned an M.S. in physics. During my subsequent job search, I was fortunate to have two opportunities: one in industry and one with NASA. Though NASA highly encouraged its researchers to earn their Ph.D. degrees, I had just finished graduate school and was determined never to return. However, I really wanted to work at NASA, so I accepted the offer. I have been pleased with that decision ever since. Within five years I completed my Ph.D. in electrical engineering and have learned "never to say never."

Since I know how challenging it is to put oneself through school, I have participated in outreach programs to help inform students and sponsored several summer interns. In addition, I championed and led the effort for a NASA-sponsor Web site (www.tech-interns.com) that offers one-stop shopping for students seeking scholarships, technical internships, and help in preparing for the job search. This site receives approximately 40,000 hits per month. I feel pleased that I have been able to make the road a little easier for those who follow.

MARGARET L. TUMA, Ph.D., NASA Glenn Research Center.

Early in my years as a NASA employee, when I was examining career opportunities and my potential next steps, I weighed the decision to pursue an advanced degree against a number of NASA and government training programs. At that time, I elected to apply to Goddard's Project Management Development Emprise, a program dedicated to the development of the next cadre of project managers, via mentoring, on-the-job training, and formal training coursework. In four years, I graduated from the program and moved toward management. Soon after, I was selected as a Fellow in the Council for Excellence in Government (CEG). The fellowship was a year long, and I was a participant in a leadership development program with over 100 other federal leaders. This experience truly was a turning point in my career; it was the beginning of significant learning and a major milestone in my journey in exploring and discovering my life's work. The year with the Council involved intensive coaching, training workshops, leadership benchmarking, and work on specific results for the year. My learning and experiential training as a CEG Fellow were invaluable and provided me with a solid foundation for my subsequent work in leadership.

KRISTI BROWN, B.Sc., Special Assistant for Strategy
and Development, Goddard Space Flight Center.

Develop Complementary Skills and Knowledge

Senior scientists at NASA learned early on that technical and scientific competence alone are insufficient to become successful. It takes mental toughness, superior communication and interpersonal skills, and access to networking and mentoring to succeed in the typical male-dominated scientific and engineering environments. Women, such as Kristi Brown, Olga Dominguez, and the author, have seen first hand that complementary skills, knowledge, and experience are essential to succeed in these environments.

KRISTI BROWN continues her story: Following my fellowship in CEG, I completed several courses within NASA and the federal sector that then led me to executive coaching. Coaching is a unique approach to learning, especially with respect to professional and personal development throughout one's entire life. Through coaching, I have gained clarity regarding my values, talents, strengths, and challenges; set lifelong goals with interim milestones; and achieved much in my own self-examination and growth. In addition to coaching, I have been deliberate in seeking mentoring from others and in pursuing a balance and diversity of experiences and training in my work to enable me to develop further through experiential learning.

Your life and your work will be enriched through building a network of supportive colleagues and friends. Reach out to others; accept and give support. Seek a mentor and/or a coach. I have been blessed with incredible role models, mentors, coaches, and teachers along my journey. It is important to recognize that your learning is most powerful from others who have traveled on paths before you.

For me, probably the most significant influence in my journey was becoming a mother. During my university years, my generation of women was taught "you can have it all." Throughout the early years of my career, I was driven by aspirations of "having it all." After motherhood, my motto became "you can have it all, but not all at once!" As a proud parent of two sons, I have truly mastered the meaning of balance in my life and have been able to place my professional work in context. Work is important but has paled in my life's context now that I have children. Five years ago, a life-altering event occurred. At 10 weeks of age, my second son became extremely sick and was diagnosed with a rare illness. But, fortunately, he survived, and today he is a delightful, normal, and extremely energetic little boy. The experience of almost losing my son had a tremendous impact on how I view my work; it has taken on a whole new meaning. While I am not with my children, my time at work has to be a valuable investment. I am driven to truly make a difference in the lives of others each and every day.

In order to affirm your work in the context of your entire life, you must invest in yourself. Examine your values in the context of your career choices. Set yourself on a path of powerful learning; your opportunities for growth and development are tremendous. Most significant to your success will be your willingness to invest in yourself. Seek the insights and help of others on your journey. It is never too late, but certainly it is never too early, to begin. Remember, life is indeed the journey, not the destination.

My 25-year career as a civil servant has spanned local, state, and federal governments. In all this time, the experiences I remember as being the best are those in which I learned about myself: how I integrate ideas, take in and process information, what are my weaknesses. It was the feedback from supervisors and peers that helped me grow as a person and as a professional. Very early in my career, I learned an important lesson that has served me well since, that is, to take an active role in promoting myself. I was in my 20s and one of the first women hired to the Water and Sewer Division of a local county office that offered a variety of services including food inspections, air quality monitoring, and water and sewer design and inspection. The work was outdoors, physically demanding, and consisted of working with construction crews, architecture and engineering firms, and developers. I learned how to read plans, shoot transits, and design water and sewer systems. Since my degree was in fish and wildlife management, with a minor in zoology, this was all new to me. Within the first year, I became one of the experts in the division. I asked lots of questions of the division senior staff and asked the contractors, developers, and members of the architecture and engineering firms to teach me how to operate the tools of their trade. I learned how to operate a backhoe, a well drilling rig, an engineer's transit, and many other construction tools. This earned me the respect of my coworkers and customers.

I enjoyed the work a great deal but was becoming frustrated professionally. I had trained several individuals who eventually became senior to me. I felt cheated and discouraged. I started job hunting and was offered a job at the State level. Before I left, I spoke with the Director of the county office, Mr. Dew, to offer my thanks and say good-bye.—My father had taught me never to burn my bridges, and I did not want to leave without saying something.—I thought the conversation would take all of five minutes, and I'd be out of there for good, hoorah! Well, 45 minutes later, I emerged, a bit dazed and with an important lesson that has helped throughout my entire career. What the Director told me was that he had heard wonderful reports about me, that I was one of his best workers, and that I was welcome back anytime. He also told me why he had never promoted me into a more senior position. He explained that if I wanted to succeed in life, I would have to "toot my own horn" because no one else ever would. He told me to take pride in what I achieved and to tell others about my own accomplishments. Having been raised as a woman in the 1950s and 1960s, this kind of self-promotion was neither natural nor comfortable. Women were expected to be "the power behind the man," home bodies, and not interested in working. Although I had wonderful parents who encouraged me to be everything I could be, society was not as encouraging and looked down on women who "bragged" about themselves. Yet, thanks to the support and understanding of my parents, I listened to what Mr. Dew had to say. I listened to his feedback, did not take it personally (after I got over my initial anger and frustration), and began to incorporate behaviors that would lead me to become more effective in promoting my abilities and experience. I stopped saying "I think" as a preface to every sentence (especially when I knew I was correct), stopped tilting my head—a nonverbal behavior indicative of submission—though I'm still consciously working on this one (old habits die hard), and I began taking credit for my work. I have offered this advice to many colleagues—both men and women—throughout my career. I don't know if my former director has any idea of the impact his words have had on me some 20+ years later or of the impact he has had on others as I pass on his advice. Even now I strongly recommend that people not take negative feedback personally. Instead I say: "analyze the situation and the comments, see the truth behind the hurt, modify your behaviors to affect change, and 'toot your own horn!'"

Olga M. Dominguez, Deputy Assistant Administrator, NASA.

Do not minimize the fact that racial and gender problems still exist. It is imperative that you know your rights and learn to cope.

At this point in my career, I feel talented, successful, and ready to move into senior management. But starting out, I was to discover that my journey would sometimes be an uphill climb. These were often the years when my efforts were mocked and my abilities ignored. At times, my experiences were traumatic. But I learned the importance of reexamining who I was, my capabilities, and my goals. A turning point in my career came in 1983, when I faced the decision of whether to stay and influence the environment or seek other opportunities. I received a telephone call from a former boss who inquired if I would be willing to consider a special assignment in Washington, DC. I said, "Yes." While in Washington, I had an opportunity to participate in a one-week course on the human element in work, and the importance of being proactive in managing your own career. I was inspired to plan for my future, stick with my goals, and call on people to help me—"The three Ps: planning, persistence, and people." I decided to take control of my career. Armed with determination and renewed confidence, I began my years of "stepping up" and have used the following guidelines for continuing professional development ever since. (1) Upgrade your skills periodically—it is your skills, talents, and accomplishments that get you to the interview. (Always schedule an exit interview if you are not selected for a job.) Learn of your weaknesses. Decide what, if anything, you can do about them; then do it. This is an important aspect of growth and maturity. (2) Call on people when you need them; they are excellent resources. (3) Stay focused on your goal. An executive coach shared with me an analogy from baseball that has served me well: Throughout your working life, people will throw balls at you (e.g., curve balls, high balls, low balls). To stay focused, respond only to those "balls" that keep you in the game you want to play (i.e., don't play by other people's rules nor respond to the agendas of others). It is acceptable to ignore feedback that will not help you reach your goal (e.g., do not spend time with naysayers or on negativity). (4) Then, find your place. Don't expect someone else to identify it for you. If you are not valued where you are, seek another place.

I learned a valuable lesson in Washington, DC. I was talented and knew I could succeed. With renewed confidence, I applied for new positions and was awarded them. At night, I attended graduate school and earned extra academic qualifications. I sought senior executive positions and earned them. Through planning, persistence, and the support of helpful people, I have found "my place in space." ANNGIENETTA R. JOHNSON, D.Sc., Assistant Associate Administrator for Education, NASA.

REFERENCES

Bundesministerium für Bildung und Forschung. *Buhlmahn: Frauenförderung ist Chefsache* (2001, 2003) [Online]. Available at: http://www.bmbf.de/press/776.php (accessed August 20, 2005).

Brouns, M. The gendered construction of scientific quality. In: *Hochschulreform, Macht, Geschlecht*, Michel, C., et al., eds. Bern: Bundesamt für Bildung und Wissenschaft, 2003, pp. 89–98.

Center of Excellence Women in Science. *Programm "Anstob zum Aufstieg"*(2004) [Online]. Available at: http:www.cews.org/cews/bertra.php?aid=73 (accessed August 20, 2005).

Conference Board of Canada. *Employability Skills 2000+*(2005) [Online]. Available at: http://www.conferenceboard.ca/education/learningtools/pdfs/esp2000.pdf (accessed August 20, 2005).

Faerber, C. Frauen auf die Lehrstühle durch Gender Mainstreaming? In: *Gender Mainstreaming: eine Innovation in der Gleichstellungspolitik*, Bothfeld, S., et al., eds. Frankfurt/New York: Campus, 2002, pp. 107–131.

Rees, T. *National Policies on Women and Science in Europe. A Report about Women and Science in 30 Countries.* Brussels: European Commission. The Helsinki Group on Women and Science, 2002.

••

MARY OSBORN, Ph.D.
Max Planck Institute for Biophysical Chemistry, GERMANY
Life Scientist

••

Professor Mary Osborn has made a career in three countries: the United Kingdom, United States, and Germany. She earned a first degree in mathematics and physics at the University of Cambridge, England, a Ph.D. in biophysics at Pennsylvania State University, and had a postdoctoral fellowship in Dr. J. D. Watson's laboratory at Harvard.

Her first staff positions were at the Laboratory of Molecular Biology in Cambridge and at Cold Spring Harbor Laboratory. Currently, she is a cell biologist at the Max Planck Institute for Biophysical Chemistry in Göttingen and an honorary professor in the medical faculty at the University of Göttingen. She holds an honorary doctorate from the Pomerian Medical Academy in Sczeczin, Poland. She won the Meyenburg prize for cancer research and the 2002 L'Oreal/UNESCO Prize given to women for excellence in science.

Her research interests have focused on the cytoskeleton, certain proteins of the cell nucleus, and the use of antibodies in cell and tumor typing. Antibodies made in her laboratory have been licensed to companies worldwide.

She was a trustee of the Swedish Foundation on the Environment, MISTRA, and has chaired both the Scientific Advisory Board of the European Molecular Biology Laboratory in Heidelberg and the Cell Biology Section of Academia Europaea. She is the current President of the International Union of Biochemistry and Molecular Biology, an organization that represents biochemists and molecular biologists in 72 countries.

She was rapporteur and speaker at the first European Commission workshop on Women in Science in 1993 and a keynote speaker at the European Union Conference on Women and Science in 1998. She was Chair of the European Technology Assessment Network (ETAN) on Women and Science, sponsored by the European Union Research Directorate, which produced the ETAN report "Promoting Excellence through Mainstreaming Gender Equality." She continues to speak out on issues related to women and science.

Chapter

3

· · · · · · · · · ·

TRAINING AND WORKING ABROAD

· · · · · · · · · · · · · · · · ·

Mary Osborn

THE ADVANTAGES OF MOBILITY

Spending a period abroad to pursue training and/or work opportunities can be very beneficial, not only to the scientists themselves but to their employers, their home countries, and society itself. One of the goals of the European Commission (2001) is to stimulate mobility of scientists in Europe across national boundaries. Advantages include:

- improving the quantity and quality of research training
- fostering international collaboration
- fostering contacts between academia and industry
- enhancing transfer of knowledge and technology
- raising the scientific excellence of individual researchers
- furthering a more equal distribution of research excellence throughout the European research area

At a symposium in 2003, DiMaggio summarized the benefits of international mobility for research and development:

- access to centers of excellence and training of young researchers
- research tends to become global and complex problem solving often requires international cooperation
- mobility is vital for idea generation/circulation
- transfer of knowledge works "people to people"
- long-standing collaborations may arise from scientific exchanges
- personal growth (scientific and cultural)

In addition, I would emphasize that international mobility provides:

- chances for young people to do research in the best labs in their field
- insight in choosing a research topic when you set up your own lab
- help with building networks, which are often vital for later success in science
- an opportunity to acquire fluency in English if you come from a non-English speaking country

Obviously, these advantages are not limited to exchanges between European Union (EU) Member States. They are equally applicable to exchanges between, for instance, The Netherlands and Australia or Italy and the United States (US). Nor are the positive aspects of a stay abroad limited to the acquisition of new knowledge in your own discipline. You also will be exposed to a new cultural environment or language and gain a new perspective on your home country. In the pages that follow, I will comment on some of these points, particularly as they apply to a career in the 21st century.

Whether a period spent abroad is essential depends in part on the country you come from, your field of research, and your career ambitions. An excellent book on career advice for life scientists written in the US (Marincola, 2002), for example, does not mention mobility, and indeed the pressures to spend a period abroad may be less for scientists from the US than for others. In other countries, the perspective is quite different. Increasing amounts of money are being invested to stimulate mobility across national boundaries. For example, the Marie Curie Program of the EU, Human Frontier Science Program (HFSP), European Molecular Biological Organization (EMBO), and other national organizations provide stipends for a stay abroad. Some countries (e.g., France) provide very little financing for postdoctoral fellowship positions in the home country, so people at this stage in their careers have little choice but to go abroad. While a stay abroad is becoming essential for natural scientists, it is less so for those, say, in the social sciences (although even here there is increasing pressure to have international experience at some stage). Career ambitions may also play a role. Stays abroad will be most important for those planning a career in academic or industrial research.

MIGRATION PATTERNS FOR SCIENTISTS

Scientists have globally marketable skills. This can be demonstrated by, for instance, mapping individual moves. In my own case, I earned my first degree in England and my Ph.D. in the US. Then I held a position in the Laboratory of Molecular Biology in Cambridge, England, before returning to Cold Spring Harbor in the US for two and a half years. In 1975, I moved to Göttingen, Germany, where I have been since. At each stage in my career and at each of the institutes or universities that I was part of, I had the opportunity to build networks among my colleagues. But it was only in 2003, when I attended a meeting at Cambridge to celebrate 50 years of DNA, did I realize how many good scientists had been through the Laboratory of Molecular Biology and through Jim Watson's laboratory at Harvard, and how important such networks are—for the daily work of research and for building a career (as you'll read in the chapters "Networking," "Transitions," and "Climbing the Ladder").

A second way to demonstrate the transferability of scientific research skills across national borders (and map international migration in science) is to identify the countries of origin of the students, postdoctoral fellows, and scientists working in a single institute at a single point in time. Thus, in my own institute (the Max Planck Institute for Biophysical Chemistry in Göttingen) in November 2002, most regions of the world were represented, with the exception of Africa and some parts of South America.

Migration patterns between countries or between regions of the world are not necessarily balanced. For example, if one looks at exchanges between candidate countries and the EU that were funded by the European Commission Research Directorate's "Fifth Framework Programme" (1998–2000), 18% of EU fellows came from candidate countries to work in EU Member States, while only 1.5% of EU fellows from EU Member States chose to work in candidate countries. In the past decades, there has been a brain drain to the US from Western Europe and from other countries. This is in part because of the excellence of US science and the favorable situation for science funding in the US, including individual salaries. However, in part it is because scientists from abroad have been made to feel welcome both in and outside the laboratory, because of the pioneer spirit in the US and the spin-off (both to basic research and to commercialization) resulting from discoveries made in basic research.

Basic research not only increases knowledge but also solves problems perceived by society as important that are too long-term for industry. It yields surprises that result in new products or processes. Examples include lasers, x-rays, semiconductors, xenography, global positioning systems, the human genome sequence, monoclonal antibodies, and the Internet (Committee on Economic Development Report, 1998). Seventy-three percent of the scientific papers cited in patents were basic research articles funded by governments or nonprofit organizations (Narin et al. 1997). For these reasons, other governments in Europe and beyond have begun to place more emphasis on research, and it may be that by the middle of the 21st century, countries such as China, Japan, or Korea may attract increasing numbers of scientists for short or long stays.

Results of the first call for Marie Curie Actions (i.e., training and mobility activities) in the EU's "Sixth Framework Programme" show some interesting features of migration patterns in and out of the EU and between EU Member States. For fellowships to come from outside the EU to the EU, the largest number of applicants are from Russia, China, India, and the US (in this order). The most popular destinations among the awardees are the United Kingdom (UK), Germany, and France. For fellowships awarded to scientists to leave the EU to go abroad, the largest number of applicants come from France, followed by Germany, Italy, and Spain. The most popular destinations among the awardees are the US, Australia, and Canada. Finally, for exchanges within the EU, the largest number of applicants are from France, followed by Spain, Italy, and Germany. For this program, the most popular destinations are the UK, France, and Germany.

In general, however, our knowledge of migration patterns for those in academe or industry is still very limited. A National Science Foundation study has shown that 71% of foreign students who completed a Ph.D. in 1999 were still in the US in 2001. The exact percentage depends on field (computer sciences, 80%; life sciences, 77%; social sciences, 55%) and on country of origin (China, 96%; UK, 53%; France, 30%) (Finn, 2001). In addition, of those who earned their Ph.D. in 1991, 58% were still in the US in 2001. Today one can locate almost any life scientist by typing her or his name into one of the literature databases to determine recent publications and the institute where the individual is located. Using this technique, EMBO undertook a study of where scientists whom they had supported as postdoctoral fellows were located 10 years later (Gannon et al., 1997). For the 1984–1985 cohort of fellows, 73% of those who could be located 10 years later had returned to their home countries. The HFSP in a direct survey of their fellows from 1990–1998 found that, in the year 2000, 44% of the fellows had returned to the home country, 41% had stayed in their host country, and 12% were in a third country. Finally the Deutsche Forschungsgemeinschaft (German Research Foundation)—the major funding organization in Germany—has determined that of the fellows whom it funded for a stay abroad, 40% applied for grants in Germany six years later.

TIMING AND FUNDING PERIODS ABROAD

There are several stages in a scientific career when thought can be given to the possibility of studying or doing research abroad. These are at the undergraduate level, the earning of a Ph.D., the postdoctoral phase, the first or subsequent job, and a sabbatical.

The amount of time spent abroad, the ways of funding it, and the benefits of international experience are very different, depending on the choices you make. The possibilities today are such that exchanges may be between virtually any two countries in the world. For scientific training, however, language may also be an important factor. If one's first language is not English, a stay in a laboratory in an English-speaking country has obvious advantages. However, most international laboratories that admit students and postdoctoral fellows from around the world today will use English as the everyday language. Almost all laboratories of international standing will publish their results in journals in English. The exception to this may be France, where there is still considerable pressure to use the French language.

If you are not sure which country to go to or what facilities are available at particular institutions, you can find a lot of useful information on the internet, both with respect to working abroad and to specific opportunities in your own specialty. Useful sites include:

- ExpatExpert (www.expatexpert.com/links) for information and advice on working abroad
- Your Europe (europa.eu.int/youreurope), a good site for EU citizens moving from one country to another
- Researchers' Mobility Portal (europa.eu.int/eracareers/index_en.cfm), a new site run by the EU designed for researchers looking for career opportunities and for relevant information and assistance (look particularly at ERA.MORE, the European Network of Mobility Centres)
- Life Sciences Mobility Portal from EMBO (mobility.embo.org/html/index.php), a portal designed to give researchers worldwide fast access to mobility across Europe
- Science NextWave (nextwave.sciencemag.org), the career development source for scientists, has interesting articles about career experiences

in different countries; US, Canada, Germany, UK, and The Netherlands have their own NextWave home pages

At which stage in a scientific career is mobility most important? Results of a recent survey on the EMBO Mobility portal are shown in Figure 3.1. Although the sample in their survey is small, most scientists whom I know would also single out the postdoctoral stage as the best time for a stay abroad.

Ackers (2001), in a study of female Marie Curie Fellows, provides an excellent overview of individual reactions to taking a fellowship abroad. In addition, she documents that young single women and men display no significant differences in willingness to move if it advances their career. Interestingly, gender-specific patterns seem to emerge at later career stages.

Undergraduate Level

At the undergraduate level, exchanges tend to be relatively short. Individual American and European universities often have specific exchange programs allowing undergraduates to study abroad for a semester or sometimes longer. Although fees still have to be paid to the home university during this time, the advantage of such specific exchange programs usually is that the credits obtained abroad are recognized without problem by the home university. Students arranging their own exchanges may have

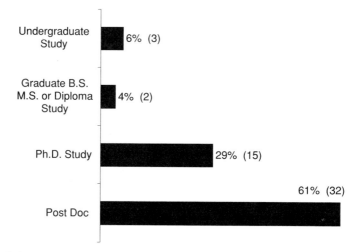

Figure 3.1   At which stage in your career would you rank mobility as most important? (Source: http://mobility.embo.org/html/index.php.)

difficulties obtaining credit for the courses that they have taken. Many universities in Europe, for instance, do not issue grades to those who study for only a semester; this is a point that needs to be discussed with your home institution before leaving. For those wanting to study at universities in the US, it is important to allow sufficient time for visa requests to be processed, even for short study periods abroad. Temporary working permits were relatively easy to obtain before the events of 9/11, but since then, restrictions have tightened.

Earning a Ph.D.

Traveling from country to country with science is part of my life. So far I have trained in three countries: Romania (where I was born and studied as an undergraduate), the Czech Republic (where I completed a four-month research project as part of my undergraduate studies), and Germany (where I am currently completing my Ph.D. as a member of the International Max Planck Research School, run jointly with the University of Göttingen). It will not stop here, as I enjoy getting to know new societies, learning new languages, and, more importantly, working in science. Through such experiences, one becomes more open minded, more tolerant of people, and more creative and efficient in one's own activities.

Gabriella Ficz, Ph.D. Candidate

After earning your first degree in your home country, you may wonder about studying abroad. Doctoral programs usually take four to five years (with certain exceptions, such as in the UK). Depending on the country or type of graduate program selected, some of this time may be spent in course work, laboratory rotations, or mini research projects. Only after this introductory time does one begin the real research project that will lead to your Ph.D. The first challenge that you face is deciding what you want to study and where. The range of research topics is much wider at this level than the choice of disciplines at the undergraduate level. Indeed, it is often possible to switch to a new discipline or to an interdisciplinary field (e.g., nanotechnology, biophysics).

The advice to go to the best possible laboratory that you can get into needs to be taken very seriously. Your choice of graduate program will determine not only the choice of laboratory where you will do your work but may also determine much of your subsequent career path. This is because during your Ph.D., you will begin to build, through your advisor and others in the department, the national and international networks that will be important for subsequent career steps. It may be difficult enough to identify the best departments or universities in your own country (there are often very strong departments in lower ranked universities and vice versa), but to find those in a foreign country is even more challenging. However, note the following:

- Some countries have a clearly established hierarchy of universities. In the US, for example, the so-called "Ivy League" universities (among them, Harvard and Yale) enjoy a particularly good reputation. In the UK, Cambridge and Oxford are particularly attractive choices.
- Other countries (particularly in Europe) may also have nonuniversity institutions that focus on research, in addition to the university laboratories. In Germany, for instance, there is the Max Planck Society, the Helmholtz Society, the Leibniz Association, and the Frauenhofer Society. While these institutes do not award degrees, the research is performed within the institute and arrangements are made with a local university for the student to earn a degree. The Pasteur Institute in Paris, the Laboratory of Molecular Biology in Cambridge, and the Karolinska Institute in Stockholm are other examples. A more recent initiative organized in Germany is international graduate schools. Currently, the Max Planck Society has 29 such schools collaborating with German universities (for details, see www.mpg.de/english/institutesProjectsFacilities/schoolChoice/index.html). These bring together graduate students from different countries, including Germany. Teaching is in English. Entry is by a combination written examination and interview process.
- International laboratories are another possibility. Examples include the European Molecular Biology Laboratory in Heidelberg, Germany (www.embl-heidelberg.de) for life scientists, and for physicists, CERN in Switzerland (public.web.cern.ch/public).

All the institutions listed will give the graduate student an excellent and very solid graduate training. Be aware that some universities (e.g., in the US) have application procedures requiring you to take examinations, such as the Graduate Research Examination or the qualification to demonstrate that you can read and write English (TOFEL), before you are offered admission. Other institutions (e.g., German and Spanish universities) may have problems determining what your previous degrees (earned in other countries) are equivalent to in their educational systems. Sometimes there are ways around such difficulties. For example, there is a clearinghouse in the US for degrees from China. Within a very short time, this agency will provide a clear description of a degree from a given Chinese university and indicate what the American equivalent is. Foreign student affairs offices in the US universities can be very helpful in solving such problems.

Financing the Ph.D. in a foreign country can be difficult, but often this is solved by the institution itself. In the US, for example, graduate students may be offered teaching assistantships and other grants, or they may be paid directly from their supervisors' research grants. These stipends usually include money to pay any fees charged by the university. At the International Graduate School in Germany, students receive a stipend from the graduate school for the first year and after that are paid with money from grants given to the supervisor for whom they are working on a research project. In other countries, such as the UK, fees for foreign students can be charged and are often higher for foreign students than for students from the home country.

Postdoctoral Phase

By the time you have earned your Ph.D., you may have decided that you would like to continue working in the particular area of your thesis topic. If not, the postdoctoral phase will give you an opportunity to explore a second field before you have to apply for grants and fund your own research. In some countries, it is common to have more than one postdoctoral position. This probably still is the most common stage in a scientific career to spend a period abroad. In some countries (particularly in Europe), it is almost mandatory to go abroad because little or no support is provided (France is a good example).

If you do decide to go abroad, the questions are the same as at the graduate student level: "What do you want to do? Where do you want to do it? And who will pay for it?" But the answers demand a bit more precision than at the graduate student level. At this level, the emphasis may be much more on what you want to do than where you want to do it. And again, you need to choose the best possible environment to do your research. This may mean moving halfway around the world or may involve moving only a short distance (along the Charles River in Cambridge, MA, for example, to attend the Massachusetts Institute of Technology after graduating from Harvard). Regardless, it is important to change laboratories at this stage and to see a second way of choosing problems and perhaps different ways of solving them.

After answering the first two questions, you'll need to identify a source of funding. The best approach is to contact the head of the laboratory where you want to work and ask her/his advice. In general, two modes of support may be possible. The first is to apply for a postdoctoral fellowship from a source in your home or host country. This is a personal fellowship given to you for the time you will be a postdoctoral fellow in a specific laboratory (usually one to three years). Alternatively, some groups or institutions have support from an institutional or individual grant, which they can offer you directly. Be realistic about your chances of winning personal fellowships. Ask each organization how many such fellowships there are, and what the success rate is (see Table 3.1 or the ETAN report). (It is worth noting that being awarded a personal fellowship from a good organization will be viewed very positively by job selection committees and grants panels.) Be aware that some organizations in your home country may only award fellowships for one year of study abroad; this usually is not long enough to accomplish something, particularly if you are starting a new project. So inquire early of your future supervisor about what the chances might be of mixed financing (i.e., payment by the home country for the first year, followed by payment by the institution in the host country for subsequent years) or of financing for the whole time of the postdoctoral stay from an institution in the host country (Table 3.1).

Table 3.1   Success rates for postdoctoral fellowships and for 5-year young investigator positions

Postdoctoral	Success rate %	% Given to women
HFSP postdoctoral fellowships (2004)	13.3	34
EMBO long-term fellowships (1996–2003)	20.3	36
Marie Curie EIF[a]	23.6	
Marie Curie OIF[a]	17.8	
Marie Curie IIF[a]	15.8	
Young investigators		
Dorothy Hodgkin fellowships (UK) (1995–1999)	5.5	93
BioFuture (Germany)[b] (1998–2003)	4.1	21
Marie Curie Excellence Grants[a,b]	10.8	
EURYI grants[b] (2004)	3.2	
EMBO Young Investigator Program[c] (2000–2003)	13.6	22.9

[a]Numbers for first call Sixth Framework Programme (2003/2004).
[b]Provide salary and generous support for a research group to young investigators.
[c]Provide relatively little research money but provide opportunities for networking on a European level.
HFSP, Human Frontier Science Program; EMBO, European Molecular Biology Organization; EIF, Exchange International Fellowships; OIF, Outgoing International Fellowships; IIF, Incoming International Fellowships; EURYI, European Young Investigator Awards.

Note: The Marie Curie program of the EU finances exchanges of postdoctoral fellows, not only between EU Member States ("Exchange International Fellowships") but also between a so-called third country (a country outside the EU) and an EU Member State ("Incoming International Fellowships") or between a EU Member State and a third country ("Outgoing International Fellowships").

The First or Subsequent Jobs

Globalization is affecting science, as it is other professions. Thus, while it may have been common 20 to 30 years ago to take your first job in your home country, scientists now realize that they have internationally marketable skills. This can be especially relevant for individuals from low- and middle-income nations who wish to gain more experience before returning home or for individuals from countries such as Spain or Australia, where more scientists seem to be trained than can be accommodated within the system (with the result that there is a net outflow of scientists from these countries).

Increasingly, jobs are advertised internationally. For some, there are specific language requirements; for others, a transition time is allowed for the new employee to become fluent in the language. The EU in particular tries to increase the international opportunities for scientists, in line with Busquin's vision of a European Research Area in which each EU Member State will devote 3% of their gross national product to science by the year 2010. In this connection, the EU Research Directorate is working on a charter to document best practice in the hiring and employment of scientists. In addition, some countries very actively work to increase their numbers of foreign-born scientists. Examples include the US and Germany. In the Max Planck Society, for example, 25% of the directors are non-Germans, and 33% were recruited directly from positions abroad.

This is not to suggest that you can assume that all vacant jobs are automatically advertised or posted on a central, easily accessible Web site. In more than one country in the EU, entry-level, and even senior scientist, positions can be filled without open advertisement. In addition, some countries award their starting positions *ad personam* through competitions that take place at intervals. Examples here include the Institut National de la Santé et de la Recherche Médicale (French Institute of Health and Medical Research) and Centre National de la Recherche Scientifique (French National Center for Scientific Research) for positions in France. These are given for life and can be moved between institutions. Age restrictions apply on some posts. In France, the age limit is increased by one year for each child, and if you have three or more children, it is abolished. For a female scientist in Italy, the age limit is also increased by one year for each child, but also by one year for her husband!

If you plan to return to your home country after working abroad for a time, it is advantageous to stay in contact with individuals who are still there (who may be able to alert you to appropriate job opportunities), and to visit once or twice a year. You need to keep yourself informed of the emerging opportunities as you near the end of your term abroad, so you can act on them in a timely fashion. In addition, let everyone in your network know that you are looking for a job. Very good advice for those establishing their own laboratories is included in a booklet entitled "Making the Right Moves: A Practical Guide to Scientific Management for Postdocs and New Faculty," from the Burroughs Wellcome Fund and the Howard Hughes Medical Institute (2001).

The Sabbatical

Different groups of scientists view science from different perspectives, even if it appears to be the same problem. It is very important to be exposed to different perspectives. Science is an international endeavor and it is much easier to meet other scientists if you move, rather than if you stayed glued to one spot. The three years I spent at the Laboratory of Molecular Biology in Cambridge, England, as a postdoctoral fellow were among the most valuable in my career. And sabbaticals in Germany, Australia, and distant parts of the US have likewise opened the door to new scientific directions.

Joan Argetsinger Steitz, Sterling Professor of Molecular
Biophysics and Biochemistry, Yale University

Sabbatical leaves are more often pursued in the US than in Europe. They give those teaching at universities the possibility (once every seven years) of going abroad for a year (or alternatively, going abroad for six months every three and a half years). Conditions vary depending on individual universities or institutions, and while the university will often pay half the salary, additional money has to be found from foundations or from other sources. Sometimes bilateral treaties provide tax relief. Faculty members who take sabbaticals agree that these provide excellent opportunities to leave many of the responsibilities at home, to have time to think about research, and make future plans for how to proceed and, at the same time, to enjoy these advantages in an environment different from their own and make new professional contacts. Taking sabbaticals probably also helps to avoid burnout.

OTHER CONSIDERATIONS

There are other considerations that need to be taken into account when considering a stay abroad. They are discussed in the following sections.

Difficulties with Free Movement of Researchers

Scientists often require a visa to study or work in the country of their choice. The rules and regulations governing the issue of visas are dependent on the country that you want to go to and also, in many cases, on your country of origin. This is a detail that needs to be investigated early in the application process, although often, as in the US, you may have to wait until the university of your choice has accepted you before you can apply for your visa. The length of time required to obtain a visa may depend on the country you come from. In addition, visa restrictions may be in force, which may prevent your partner from entering the country of your choice (if you are not married) or may prevent your partner from working once you arrive. You may also wish to choose a country in which you can both earn the Ph.D. and work afterward, using the degree that you have obtained as a qualification. Here again, possibilities differ between individual countries.

Other Financial Issues

Among the financial issues to consider when planning a time abroad are the implications for pensions. Past the postdoctoral stage (or even during it), contributions to pension schemes may not be transferable. There may also be minimum times during which you have to contribute before you derive any benefit. In some countries, benefits may be paid out as a cash sum if you leave the country (up to seven years in Germany, for example). Similar arrangements may apply for social security payments except that payments are usually nonrefundable. In the US, for example, payments have to be made for a minimum of 10 years before you can benefit at retirement. Regulations differ from country to country. In the European Research Area, there currently is no such thing as a transferable pension, though it would be very advantageous for the mobility of scientists within Europe if it were implemented. Many academic institutions in the US contribute to the Teachers Insurance and Annuity Association—College Retirement Equities Fund, so a change of institution within that country is usually possible without loss of pension rights. These may not be issues of great concern if you plan only a two- or three-year stay abroad, but if you work in several countries over a long time period, they will become more important.

Building equity in your home country can also be more problematic (e.g., purchasing an apartment or a house). Gaining experience abroad usually delays such investments.

Personal Considerations

The idea that you can market your skills across national boundaries may take some getting used to. It requires some boldness to send off an application to a foreign country for training in a laboratory or work in a job. The first application may be unsuccessful, so it is important to plan your strategy and be persistent. If, for example, you are applying for a postdoctoral position, a useful approach is to make a list of four to eight laboratories of interest, then send applications to, say, the top five.

Take advantage of every opportunity to visit the laboratories before making your decision. If, for example, a key scientific conference in your field is being held in the same community, make arrangements to visit the laboratory while you are there. (It need not cost a fortune, given the availability of discount airlines and special fares.) You will find out very quickly whether you would fit into the intellectual and social life and culture of the group. You may find several nationalities represented among the students and scientists, become aware of the cross-cultural differences influencing interactions, and learn about the working habits and leisure interests of the members. Through casual conversations, you will learn whether people are happy with the opportunities available to them and gain a better sense of whether the work–life balance in the laboratory is appropriate for you. The information you obtain from a personal visit is rarely available from afar and usually makes your decision much easier.

The importance of personal considerations—especially how welcome you would feel and how well you would fit into the new environment—cannot be overstated. One story told at a recent symposium in Rome illustrates this. A particular laboratory in France could not understand why it had received so many applications from one area in China. It turned out that the experience of the first postdoctoral fellow from China was so positive that he had written to his previous institute and to his friends in China describing this experience. Not only had his hosts found an apartment for him, but they had equipped it with chopsticks, rice, and a few other essentials, so that he felt welcome from the moment he arrived. All of us who have spent time abroad have similarly happy memories of invitations to people's houses, excursions, or dinners in which the whole group took part.

Family Issues

It is relatively easy to move a single individual hundreds or thousands of kilometers. It is more challenging if there are two of you and if children are involved.

Child care facilities differ widely between countries. As an example, see Figure 3.2, which shows the number of state-supported child care places for children under the age of three in the UK and selected countries in Europe. In addition, school systems vary from country to country. In Germany, schools that teach for half a day are still the norm, and some schools send the students home if the teacher is ill. In many countries, care of preschool-age children has to be paid for, often on a sliding scale (with subsidies for those who cannot afford to pay the full amount). In countries with fixed charges for crib or kindergarten places, there may be assistance for those who cannot afford to pay; some educational institutions provide day care for the young children of their employees.

Sometimes it can be difficult to convince parents or other relatives that a spell abroad will benefit one's career. However, there are many examples in which parents choose to visit children and grandchildren during their

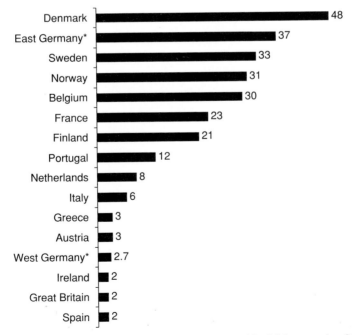

Figure 3.2   State-supported child care places per 100 children under 3 years of age. (Source: Berlin Institute for Social Research, 1997 and *2002.)

stay abroad and so have an opportunity to see countries that they would not otherwise have visited. Elder care still represents a significant challenge to mobility.

Pets also need to be considered when contemplating a move across national boundaries. It still is difficult, for example, to take dogs or cats into the UK without long and expensive quarantine periods.

Dual-Career Couples

Dual-career couples have the added challenge of needing not one postdoctoral post or research position, but two. Obviously, the preference would be to obtain two jobs in the same city. Sometimes this can be achieved by working at two universities in the same city (e.g., Harvard and MIT in Cambridge, MA) or even at the same university. Though employers are beginning to address the issue (see below), the problem is particularly acute for women scientists with spouses. There are at least two reasons for this, revealed in the now classic study by McNeil and Sher (1999) on physicists in the US. First, women physicists were more likely than men to be married to a partner who was also in physics. Approximately 45% of married female physicists were married to male physicists, whereas only 5% of married male physicists had a female physicist as spouse. Second, the women were, on average, younger (by two years or so) than their male partners. Thus, it usually is the woman whose career is less advanced at the time that a move is made and who becomes the "trailing spouse" with the disadvantages that this can entail.

Various solutions exist, including the following:

- shared or split positions
- spousal hiring programs
- alternative positions (academic)
- alternative positions (nonacademic)
- commuting
- legal responses

The appropriateness of any one approach depends on many factors, including the personal and professional circumstances of the dual-career couple, their goals for their time abroad, the employment opportunities in the country and community that they hope to move to, and the policies of the potential employers (the chapter "Balancing Professional and Personal Life" has an interesting story). At the same time, dual-career couples usually understand that they need to be much more flexible about where they work—at home or abroad—than do individual scientists.

Though institutions in some countries still have nepotism rules prohibiting spouses from holding faculty positions in the same department or institution, in general the situation is improving (US universities in particular are actively addressing the issue). More institutions are understanding that the way to hire a key individual may be to find an interesting job for the spouse; some have even set up special offices for this very purpose.

The situation in industry differs from firm to firm. For example, Partnerjob.com (www.partnerjob.com) is an excellent example of how a scheme can be set up to benefit both individuals and companies at an international level. Other examples include NetExpat based in Brussels (www.netexpat.com) and the Shell Spouses Employment Scheme based in the Hague (www.shellspouseemployment.com).

The most common reason for an assignment abroad to fail is partner dissatisfaction. Undoubtedly, some dissatisfaction is attributable to a spouse having to give up a well-paying job in the home country. In many instances, visa regulations may allow one partner to work but impose greater restrictions on the working permit for the spouse. Again, these are details that need to be thought through before making a final decision. Excellent descriptions of how different dual career couples resolve these problems can be found in McNeil and Sher (1999) and on the Science NextWave Web site (nextwave.sciencemag.org; use search engine to identify articles on dual-career couples).

RETURN AND CAREER DEVELOPMENT

After completing your graduate studies or a postdoctoral stay abroad, you will be facing another change: to another position abroad or a return home. With a new Ph.D., for example, you may choose to complete your postdoctoral work in a second foreign country. Or perhaps you were fortunate enough in your postdoctoral phase to have won a fellowship (e.g., some Marie Curie and HFSP fellowships) that included funds to support you for a year after your return to your home country to help you become reestablished. (The chapter "Balancing Professional and Personal Life" has a story about this.) A second way to fund your return is to apply for a grant that allows you to set up an independent five-year group in your home country, such as the Marie Curie Excellence Grants (EU), the European Young Investigator Awards Program, and the BioFuture Program of the Bundesministerium für Bildung und Forschung (Federal Ministry of Education and Research, Germany). All three programs provide generous funding for five years and allow young scientists to build up their own research program while bypassing more conventional career steps. Another option for returning home is to seek employment in a university or nonuniversity institution.

As mentioned earlier, the chances of obtaining a tenure track position differ between countries and between disciplines. One situation that you do need to avoid is becoming a "perpetual postdoc," that is, moving between laboratories or between countries every two to three years for temporary positions. For anyone wishing to climb the career ladder in science (see the chapter "Climbing the Ladder" for a complete discussion), getting a foot on the career ladder in one system is an important first step. Rotating between jobs and countries for too long is a definite disadvantage.

The longer you have been abroad, the more difficult it may be to return. If your scientific ties to your home country have decreased during your absence, you will be less aware of emerging opportunities at home. Also, you may have come to feel more "at home" in the host country and therefore less motivated to return. Again, these decisions are personal ones that may become more complicated if there are children involved and especially if the children are approaching the age when they enter first grade.

A decision to stay in a foreign country is not irreversible. In an academic career, for example, reentry is possible at several levels: after the Ph.D., after the postdoctoral phase, after some years as a group leader, or at the top, as a full professor or as head of a group in a nonuniversity institution. Also, in Europe we are seeing examples of excellent scientists who do not wish to retire at 65 or 68 and who are therefore choosing to move to countries with less stringent retirement policies. International mobility is increasing and is accepted as a fact of life in science. Indeed it is one of the advantages of a scientific career.

RESEARCH AND DISCOVERY AT AN INTERNATIONAL LEVEL

Collaborations in science are an enjoyable and very effective way of solving problems. Other scientists often have skills that are complementary to those of your own laboratory and therefore much can be accomplished by a short-term visit to another laboratory. Short-term visits can facilitate the learning of a new technique, exchange of reagents, or brainstorming to solve a particular problem, or provide the opportunity to write a joint grant.

There are a variety of ways to fund short-term visits between countries, for example, at the EU level or via organizations such as EMBO or the Federation of European Biochemical Societies. An increasing number of agencies and organizations in North America and Europe that fund research are requiring applicants to form a consortium of scientists from different countries. Examples include the integrated project or networks of excellence at the EU level (some of which involve 20 or more groups from a variety of countries), the HFSP grants, and North Atlantic Treaty Organization grants. To take advantage of such possibilities, it is necessary to have net-works of international contacts.

Gaining recognition for one's work does not occur automatically. It involves both doing excellent research and communicating your results to others (as discussed in the chapter "Communicating Science"). Only the most spectacular papers will become known without effort; most require publicity. This can take the form of describing the results at meet-ings (e.g., in poster sessions) or in more formal presentations such as invited talks. Early in a scientific career and indeed throughout, it is impor-tant to attend scientific meetings, particularly those in your own immediate scientific field. It is the individuals whom you meet at these meetings who will supplement and extend your existing networks of contacts. Thus, meet-ings need to be viewed, not only as opportunities to extend your knowledge, but also as excellent opportunities to meet other scientists interested in the same topics. Find out from those above you which meeting in your field is appropriate and try to go to it every year. In this way, you will soon accu-mulate an ever-increasing circle of colleagues and friends (the chapter "Networking" offers other strategies for developing your network). It also is worth pointing out that scientific meetings often are held in very pleasant locations around the world; an added benefit of a career in science is that this travel may be paid for by someone else (e.g., by your employer, organ-izers of the meeting [if you are an invited speaker], or through your own research grants).

> *Mobility has played an important role in my career. My moves between the UK, Germany, and the US have broadened my horizons tremen-dously. It has been both fascinating and informative to see how science is approached in different countries. And I have forged a broad network of colleagues from all over the world with whom I still interact. My move from the European Molecular Biology Laboratory in Germany to a biotechnology company in the United States represented mobility of a different type: from pure to applied research. My time in industry allowed me to see, first-hand, the steps required to develop a new cancer therapeutic, from target validation through drug discovery to clinical trials. This has had a lasting impact on my research focus, now that I have returned to academia.*
> Sara A. Courtneidge, Ph.D., Distinguished Scientific Investigator, Van Andel
> Research Institute, Grand Rapids, MI

Spending a period abroad to pursue training and/or work opportunities can be a very positive and beneficial experience for a scientist, both pro-fessionally and personally, regardless of your stage of career (though the circumstances, timing, and duration will differ). When you live and work outside your home country, you become more conscious of science as an international endeavor and of the unique possibilities that this career provides for making international contacts and developing professional and personal relationships with colleagues from a wide variety of countries. Although individuals move on and change location, the networks that you create continue to exist and provide a long-term basis for international collaborations and for obtaining grants at an international level. Not only do you learn much about the host country, by living abroad you also view your own country from a different perspective. For most scientists, this is a decision that they do not regret having undertaken. By planning carefully and choosing wisely, you too may broaden and deepen your professional and personal life through training or working abroad.

REFERENCES

Ackers, L. *The Participation of Women Researchers in the TMR Marie Curie Fellowships.* Luxembourg: Office of Official Publications of the European Communities, 2001 (also available online: www.mariecurie.org/src/press/ackers.pdf).

Committee on Economic Development. *America's Basic Research. Prosperity Through Discovery* (1998) [Online]. Available at: http://www.ced.org/projects/basic.shtml (accessed August 20, 2005).

DiMaggio, G.M. A transatlantic R&D career: opportunities and challenges. Talk at the Descartes Prize Ceremony, Rome, 2003.

European Commission. *Communication from the Commission of the Council and European Parliament: a mobility strategy for the European Research Area.* Brussels: COM, 2001, p. 331.

ETAN Report on Women and Science. *Science Policies in the European Union: Promoting Excellence Through Mainstreaming Gender Equality.* Brussels, European Commission, 2000 (also available online: www.cordis.lu/improving/women/documents.htm).

Finn, M.G. (2001) *Stay Rates of Foreign Doctorate Recipients from US Universities, 2001.* Oak Ridge, TN: Oak Ridge Institute for Science and Education, 2003 (also available online: www.orau.gov/orise/pubs.htm).

Gannon, F., Norman, J., Kriis, M., Walter, A., and Breimer, L. EMBO fellows go home. *Nature* 1997;388:416.

Howard Hughes Medical Institute. *Making the Right Moves: A Practical Guide to Scientific Management for Postdocs and New Faculty* (2005) [Online]. Available at: http://www.hhmi.org/grants/office/graduate/labmanagement.html (accessed August 20, 2005).

Marincola, E., ed. *Career Advice for Life Scientists.* Bethesda, MD: American Society for Cell Biology: Women in Cell Biology, 2002.

McNeil, L., and Sher, M. The Dual-Career-Couple Problem. *Physics Today* 1999;52:32–37 (see also physics.wm.edu/~sher/survey.html)

Narin, F., Hamilton, K.S., and Olivastro, D. The increasing linkage between U.S. technology and public science. *Research Policy* 1997;26:317–330.

••

ILENE. J. BUSCH-VISHNIAC, Ph.D.
Professor, Whiting School of Engineering
Johns Hopkins University, UNITED STATES
Mechanical Engineer

••

A widely respected scientist and dynamic and inspiring leader, Dr. Ilene Busch-Vishniac's passion for excellence in engineering research and education has been a positive driving force in every context in which she has worked. Her success in advancing all aspects of her career, while enjoying a fulfilling personal and family life, is a tribute to her vision, commitment, and wonderful sense of humor.

An early interest in music inspired Dr. Busch-Vishniac to pursue a career as a pianist. But a first-year course on the physics of music at the University of Rochester so captivated her imagination and intellectual curiosity that she immersed herself in the sciences and graduated with degrees in physics and mathematics. Her fascination with acoustics and the engineering problems associated with sound led her to the Massachusetts Institute of Technology, where she earned her master's and doctorate degrees in mechanical engineering. After two years at Bell Laboratories in the Acoustics Research Department, she joined the faculty at the University of Texas, Austin, and for nearly two decades dedicated herself to research, teaching, and mentoring. Her excellence was recognized by the Society of Women Engineers in 1997, when they presented her with their highest honor: the "Achievement Award."

The following year, Dr. Busch-Vishniac was appointed Dean of Engineering at Johns Hopkins University—one of the very few female deans of engineering in the United States. There, she continued to distinguish herself. During her tenure, she was named the Lemelson-MIT Program Inventor of the Week (2001) and awarded the Silver Medal in Engineering Acoustics (Acoustical Society of America, 2001). The Whiting School of Engineering was ranked one of the top engineering schools in the country (*U.S. News & World Report*).

She resigned her Deanship in 2003 to take up the Presidency of the Acoustical Society of America.

Chapter

4

....

CLIMBING THE LADDER

....

Ilene J. Busch-Vishniac

Regardless of what some popular movies might have you believe, virtually everyone who attains a high-level position spends years acquiring the experience and skills that job requires, by working in lower level positions. We refer to the process of professional growth and advancement as "climbing the ladder."

As a metaphor for advancing in your career, climbing the ladder has a few problems if taken too literally. Ladders are as wide at the bottom as at the top, have evenly spaced rungs (steps), and don't permit much wiggle room. Fortunately, careers don't share these characteristics. Generally, the number of people at each level (step) varies more like a pyramid, with fewer peers as one's responsibilities (and rank) increase. Further, some promotions are relatively easy to achieve and others very difficult, so the size of the steps varies. And, although there are some paths straight to the top, advancement systems permit more individual variation than the ladder analogy suggests.

In the science fields, there are two ladders rather than one. One ladder consists of jobs that are predominantly or always technical, at the top of which might be a corporate fellow or an endowed chair. The other introduces supervisory and managerial roles, such as serving as a project leader, department head, or dean. There is significant overlap in the paths

61

represented by these two ladders, and for a good bit of your career, it is possible to move from one to the other.

WHY CLIMB THE LADDER?

While many people have perfectly successful and happy careers without ever giving thought to their career path, being deliberate about your career increases the probability that you will achieve your goals and be in positions to take advantage of opportunities. In this context, a good place to start thinking about your career as a science professional is to ask yourself whether you wish to climb the career ladder, and, if so, how far you might want to go. Do not view the answer as cast in stone—it should be reevaluated frequently, with the understanding that a change in desire might inspire different choices and actions.

There are many reasons for wishing to climb the ladder. Perhaps the job you've always wanted requires you to work your way up. Maybe you become bored easily, thrive on change, and seek the challenges inherent in assuming positions with greater responsibility. Perhaps the perks associated with advancement drive you forward or maybe the desire for greater control and impact propels you.

There also are reasons not to seek advancement or at least to wish to stop after climbing only a bit. For instance, you may have highly valued commitments outside of science that occupy a great deal of your time (family, religious organization, sports, etc.). Or perhaps you've found the perfect job for you and can't imagine making the sacrifices that would be required in moving up. In academia, for example, many professors actively avoid serving as a department head because they don't want to get involved in academic politics or sacrifice time in the classroom or laboratory.

In general, there are advantages and disadvantages associated with climbing the ladder. As one advances, the rewards and perquisites pile on. Salaries and benefits increase, offices become larger, more staff serve your needs, travel accommodations improve, and your ability to make or sway high-level decisions increases. For scientists, advancement might engender a bigger laboratory, more extensive equipment, more support for technical staff, reduced teaching responsibilities (for professors), more independence in project choice, and better parking spots. For those in technical administration, the perks might include a large salary boost, greater visibility in the organization, more discretionary funding to support projects you deem important, and a national or international platform from which to speak to issues that you consider important.

Of course, as one climbs the ladder there are also some personal challenges. In general, each advancement brings added responsibilities and decreased flexibility. Your time becomes more dictated by the needs of those reporting to you and to whom you report, particularly if you move into managerial positions. Further, while some revel in the high-profile life and increased job pressure, others shy away from the associated lack of privacy.

Each career is unique; there is no single path that is "right" for every scientist (as demonstrated through the stories in the chapters "Career Management" and "Transitions"). The key to making the best choices is to be honest about your personal and professional qualities and goals, to periodically weigh these—and your satisfaction with your current job—against opportunities that arise, and make your decisions deliberately. This process of evaluation can be exhausting, so it probably isn't something you would do more than once every few years.

HOW TO CLIMB THE LADDER

Now let's imagine that you've decided you want to climb the ladder and advance as far as possible in your career. What are some of the things you can do to facilitate this process?

Understand the Culture of the Organization and What It Values

One of the most important aspects of scientific positions is their variability. Every company, institution, and agency has its own culture. A job at General Electric is not the same as one at General Motors, and both of these differ significantly from academic positions and government jobs. To advance with your current employer, you must first understand the culture and values of your organization. Determine what matters most at your company or institution or agency and then seek to be recognized for excellence in that area. If, for instance, your university prides itself on research and emphasizes this at promotion time, then it would not be an easy place to earn advancement through stellar teaching. If your company specializes in products that beat the competition in cost, then focusing on new bells and whistles would be ill-advised.

In many organizations, the culture is not clearly defined in documents. Instead, employees learn through observation and from more experienced colleagues what the organization values and rewards. For young professionals and those new to the organization, it is particularly important to understand the expectations of the culture, for it provides guidance on what activities to engage in, and when. For instance, it will give some indication of the level of professional society involvement encouraged and when to engage in long-term projects versus less risky short-term projects.

Become Recognized

Just as organizations have a unique culture, so too science has a culture distinct from other disciplines, and the values promoted by science need to be understood if you are to be successful. In general, a good way to become recognized as a scientist is to choose carefully a topic of interest and importance and to focus all your efforts there—to become the world's leading authority in that area. This approach, as opposed to becoming somewhat knowledgeable in a wide range of areas, tends to increase your value and visibility. Others will seek your assistance when expertise in your area is needed. For academics getting started, the pressure to find research funding is intense, and there is a great temptation to respond to every funding opportunity. However, this scattershot method rarely works because the experts with experience and a reputation in a given area are the people normally funded.

Ironically, while science rewards specialization and technical depth rather than breadth, the positions that relate to administration and management of science require at least a perfunctory understanding of a broad range of technical disciplines. For instance, the leader of a product team needs to understand the abilities of each of the team members and to have a general sense of what scientific issues need to be addressed in product design. As a result, the typical career path for those desiring to advance into administrative jobs is to focus first in a particular area and to become well-known and respected in that area. Subsequently, one broadens one's knowledge base through participation in projects involving other areas of expertise, with intentional reading and questioning to gain some understanding of these other areas.

Science also tends to reward quality over quantity and impose high penalties for errors. It is generally forgivable to publish fewer articles than your peers if each of your articles is better cited. However, publishing an article with unsubstantiated claims or shoddy reasoning is not likely to go unnoticed and may well haunt you the rest of your career. It is important, therefore, to avoid the temptation to produce your work product (paper, design, standard, etc.) prematurely.

In addition to matching your approach to the cultures of science and your place of work, there are many other things that you can do to help advance your career. These include finding mentors, building a network, earning a reputation as a "doer," getting credit for your ideas, and enhancing your visibility through attending conferences and workshops.

Find a Mentor

Perhaps the most important action you can take to help your career advance is to find one or more mentors. A mentor is a person with knowledge that they are willing to share with you on some aspect of professional success. Mentors usually establish close personal ties with their protégés and develop a strong interest in seeing them succeed; they advocate for you as you make your way through the process of learning and growing. Mentors open doors and give a protégé a distinct advantage over someone without mentors. There is no limit to the number of people who can serve as your mentor. Typically, different mentors help with different aspects of your professional life. (The chapter "Mentoring" explores the topic in greater detail.)

While some employers recognize the value of mentors and have formal mentorship programs, most scientific organizations leave it to newcomers to find a mentor on their own. If this is the case for you, it's important to have some sense of who would serve you well as a mentor: someone who is knowledgeable and well respected and who seems like a good match temperamentally. Once this person is identified, it is a good idea to ask them whether they could help you with a question or two. Their responses will give you a good measure of their willingness to serve as a mentor and their ability. If you are happy with the results, then when you thank them for their assistance, you might ask whether they'd be willing to continue mentoring you. If you find the results of the first interaction unsatisfactory, consider whether a more suitable mentor is available.

Develop a Network

One of the things a mentor will do is help you to become connected with others and develop a network. Your network is that group of people you can call on for answers to questions, favors, or personal support. Through it, you can quickly become aware of decisions affecting you. Members of networks share information, so having a good one can be extremely useful. To the scientist interested in advancing in her career, a network can provide early notice of opportunities, advice, and inside information on a new organization.

Forming a network requires introducing yourself to people and keeping in touch with them, even if the contact is very infrequent and impersonal (as in an e-mail intended to provide an update). What is critical to understand is that a key characteristic of networks is that they are only effective when the relationships are predicated on *trust*. While it is fine to take advantage of networks for personal gain, it is never acceptable to abuse the trust of those in your network, for instance, by revealing particularly candid and controversial statements of others. Also, you must be willing to help others in your network if you want them to be willing to assist you at some time. (For a more complete treatment of the topic, refer to the chapter "Networking.")

Establish a Reputation as a "Doer"

In addition to finding a mentor and building a network, it is very useful for you to establish a reputation as someone who gets things done. Regardless of whether your job is in business, government, or academe, those who work near you must be able to count on you to contribute your fair share and to deliver on promises. The view that your colleagues have of you is enhanced by your commitment to do what you promise and to occasionally take on more than the minimum requirement of the job. (To those in charge, these characteristics also mark you as a good person to advance.) An important point here is to avoid overcommitting yourself to the extent that tasks don't get done. It is better to decline a role because you are burdened with other activities than to force the reassignment of a task to someone else at the last minute because you cannot meet the deadline.

Choose Your Commitments Wisely

There is another danger associated with taking on more than the absolute minimum. Depending on the specific roles accepted and the difference in your activity level compared to the norm for your position, it is possible to end up being taken advantage of and having your political clout eroded rather than enhanced. When possible, it is best to accept roles that are valued by your institution and to be in charge rather than be the person behind the scene getting things done. It is also important to make sure that others in your organization contribute as well, so you are not the first one called on every time something comes up. Otherwise, the core mission of your job may be compromised.

There is something to be said for taking on tasks in which you follow someone who wasn't very good at the job. The expectations of you will be set relatively low and your chance of exceeding them is great. In other words, the probability of people noticing a positive difference is very high.

Get Credit for Your Ideas

A particular issue that comes up for women and underrepresented minorities is the problem of getting credit for ideas. Unfortunately, it is rather common for ideas—be they technical or related to some administrative or organizational issue—to be credited to someone other than the originator, particularly if the originator is neither part of the majority nor a long-standing member of the group. A typical scenario might involve voicing an idea at some forum, only to have it meet with little or no reaction until someone else says essentially the same thing later.

There are a few ways that you might handle this sort of situation—what strategy works best will depend on your place of employment, your relationship with your colleagues, and your comfort level with the various options. The key is to stake a claim to the idea (if it is important enough to warrant doing so) with grace and good humor rather than complain of unfairness. This might be done by publicly thanking your colleague for embellishing and improving on the idea that you stated earlier. Depending on the people involved, it is also possible to correct attributions of ideas by speaking with the person in charge (of the meeting or project, for example) or with the person to whom your idea was erroneously attributed. These sorts of battles must be chosen carefully and handled delicately. There is nothing to gain and much to lose by being labeled as too aggressive or "bitchy."

Establish Your Reputation

A key part of moving up the ladder is to become recognized; even internal promotions are made easier by having a strong reputation outside your organization. Being a homebody who shuns traveling to conferences, giving seminars, or participating in high-level meetings away from the office makes it very tough to build a reputation, particularly for those starting out in their scientific careers. It is important for your name to be recognized and your work known and for you to generally engage in some acceptable public relations work about your contributions. For those on a tight budget this means choosing your opportunities wisely so you have the best exposure. A good rule of thumb for research work is to publish in the best journal, but give a conference talk at a meeting run by a different organization from the one publishing your paper. For people traveling for a business, it is far better to see a big customer than a little one. Further, it is a good idea to keep people informed of your successes (without overdoing it). Most organizations engage in public relations in the form of internal or external press releases, and to be the subject of such a release significantly raises your chances of climbing.

Another way of enhancing your chances of gaining recognition is to take some calculated risks in the topics you pursue. While there is clearly nothing wrong with being the person to work on the next incremental improvement in some area, you are far more likely to earn attention if you move the field in a new direction. The downsides of such an approach are significant: There is less of a base from which to build, so you are likely to make more false starts, and the risk of the research failing is much greater. You also are likely to encounter resistance from those entrenched in the field, for they have a vested interest in maintaining the status quo. On the other hand, successes in taking a new approach to a technical problem will gain you notice quickly. The advice of some senior scientists is for junior scientists to pursue a safe route in their work; the advice of others is to take some risks to set you apart from the crowd. A middle ground is to have a range of project types in your portfolio, with some that are conventional and safe and others that have a greater risk of failure but also a greater potential for impact if they succeed. For the latter projects, it is very important to work to minimize the downside risks.

Finally, there are two actions that are very useful at the point that you get serious and deliberate about moving up a notch: moving to move up and doing your homework. First, let's consider the issue of moving from your current place of employment.

Relocate for Advancement

Most scientists think about climbing the ladder where they started their career. After all, moving a laboratory and starting over is very disruptive, professionally and personally. And in their current institution, they know the people and culture and (hopefully) are happy there, so it's comfortable. However, there are some good reasons to consider relocating to a new place as you advance. By considering moving to a new employer, you open a huge range of new possibilities and increase the chances of finding a good match for yourself at the next level. In most cases, the fact that you are being courted by someone else increases your perceived value significantly at your current location. Also, by moving to a new employer, you can begin anew, with a clean slate; "sins" in your past are unknown. For administrative and managerial positions, such a move not only gives you a fresh start, but affords you a "honeymoon" period during which you can often set a new direction for the organization with less than the normal resistance to change. Finally, a move gives you leverage that you wouldn't likely have in an internal promotion because the new employer knows that they need to woo you away from your home. That wooing process, if handled well, can result in important parameters for the new job that will raise the probability of success by whatever measures you set. For instance, a move might mean more laboratory space, equipment, support staff, discretionary budget, or flexible working hours. Everything, absolutely everything, is negotiable in a move to a new place of employment.

Do Your "Homework"

In some workplaces (depending on the culture), there is a great deal of public disdain expressed for those promoted beyond a certain level. In academia, for example, most faculty express a desire to avoid being department chair, dean, provost, etc. In some companies and research laboratories, the same sort of contempt of high-level positions exists. However, to carry this attitude to a job interview is a mistake, albeit one that is often made. If you decide to pursue a promotion, do so seriously and with the same care that you use in your scientific work. Do your "homework" and learn about the expectations of the job, the performance of the current position holder, and the circumstances surrounding his or her departure, the reporting structure, and the sorts of people to whom you'd be reporting. If it's a position with a new employer, find out everything you can about the organization's financial health, its record on diversity and successes for new hires, and its culture. All this research will enable you to ask very thoughtful questions at the interviews (which always has a positive impact on your chances of success) and help you to determine the extent to which the promotion is a good fit.

DECIDING IF AN OPPORTUNITY IS RIGHT FOR YOU

Now let's assume that you have decided to pursue new opportunities and at least one option for a new position exists. Let's also assume that it is not an internal promotion for which you'd be considered automatically (such as a faculty promotion with tenure) and the option to remain in your current position exists. How do you decide if this opportunity is one that you will seize?

Deciding which opportunities are worth pursuing is an individual matter, but there are questions you can ask yourself whose answers will help you make a rational decision. These questions are only useful if you answer them honestly and permit yourself the chance to change your mind from time to time about your career goals.

Very often, job changes are made at the point someone has become frustrated or unhappy in their position. Usually, this has built up over an extended period of time and leads to a decision to move. Unfortunately, such a situation is not optimum for climbing the ladder because the motivation to grab the first opportunity that comes along is very high. Instead, it makes sense to try to be deliberate about career moves and to let current job satisfaction play a role without being the only deciding factor.

Being deliberate about your career means taking the time to think about what your dream job would be. This should be *your* dream job and not necessarily that which others hold up as the ultimate job for a scientist. Further, although it's nice to have a goal in mind, it's important to be flexible enough to revise your goals as you mature and your desires change. Being deliberate about your career implies taking steps appropriate to getting that dream job before you retire.

The most important question to ask yourself when presented with an opportunity is whether it will move you significantly closer to that dream job than you are now. If the answer is "yes," then it's worth considering. If not, it probably isn't worth pursuing. Getting closer to your dream job does not necessarily mean a promotion or climbing the ladder in a traditional sense. Sometimes, a lateral move is strategically good because it will bring you new experiences and perspectives that will be useful as your career advances.

Another question you might ask yourself, particularly at relatively early stages in your career, is whether the move is likely to eliminate options that, later on, you might want to reconsider. For instance, a sales job in a technical business is unlikely to preserve your ability to pursue research. A short-term assignment with government or a nonprofit organization, on the other hand, may enhance your ability to move up by enabling you to develop valuable contacts and experience. Eliminating options for the future is not necessarily unwise, if you are confident you know what you want.

A third question to ask yourself is whether the new job is something that you believe you will do well and enjoy. To answer this, you need to understand the nature of the work as well as the habits of the people to whom you will report. What are your instincts telling you? Will this be fun? Sometimes you won't have the answers. You may be drawn to the new job, for example, because it offers you a chance to learn new skills; you may have no idea how well you will perform.

Finally, you might ask whether this new opportunity is for a job that you can imagine yourself doing for an extended period of time. Even if it is a position that you know you would enjoy in the short term, one that could be used as a springboard for bigger and better opportunities, do you think you would be content if your career stalled at this particular point?

Putting the answers to these questions into your personal context, including the impact on all the nonprofessional aspects of your life, should make it somewhat easier to decide whether an opportunity is worth pursuing—although it always is a matter of deciding with imperfect and incomplete information. Once the decision is made, the temptation to revisit it can be very strong. But there are a few things to remember that make putting it behind you easier. First, it normally is the case that, where there is one job opportunity, there are many. Saying "no" to one usually does not mean that there never will be other options. Also, even if you choose to remain where you are, the fact that someone else was serious about trying to recruit you tends to greatly enhance the view of your value where you are.

This can be used to improve your current situation, within limits. Additionally, by changing positions at appropriate intervals, employers will not begin to wonder whether it is worth the effort to bring you into their organization. The amount of time that you should commit to a particular job while declining all other options depends on the field, your seniority, and your personal situation. In computer fields, for example, it is not unusual for people to change companies every other year. For those most junior in industry, a job change after only a year or two is also the norm rather than the exception. At more senior levels, on the other hand, changes are fewer and farther apart.

Personal Traits that Help

Scientists like to think of their profession as the last true meritocracy: If one is a skilled, productive scientist, rewards will follow. In reality, this is no truer for scientists than for any other professionals. Personal characteristics matter and play a significant role in advancement. They affect whether colleagues want to be associated with us, want us to represent them, and see us as admirable.

It is impossible to overstate the importance of treating people well in your professional life and gaining a reputation as one who does so. Treating people well doesn't necessarily mean paying employees more than your competitors would or delivering large gifts to everyone at the year's end, although these can't hurt. It means making it clear that all your professional contacts—staff, colleagues, vendors, students, janitors, all—are valued by you for the service that they perform. You demonstrate this by taking the time to get to know the people whom you meet regularly, by taking extra steps to support those who are deserving (nominating them for staff, student, or professional awards, for example), by showing compassion and patience (within reason) to those who are having a rough time, by keeping confidences, and by always being fair in the distribution of resources and judgments. By contrast, the scientist who regularly treats colleagues and staff with disrespect may well have problems gaining promotion since she will not have earned the support she needs.

Advancement as a scientist normally means an increase in the number of people who report to you, regardless of whether you choose a more administrative career or one in pure research. Thus, advancement means that proportionately more time is spent on issues that generally fall under the heading of "personnel management": hiring, advising, promoting, evaluating, and sometimes firing people. It simply is not possible to be effective in these roles without having the respect of the people who report to you, so developing good interpersonal skills is essential.

An impeccable reputation for integrity is also essential for moving up. Scientists may think they work alone, but all of us rely on the honesty of our colleagues in reporting results and, closer to home, on the word of our colleagues since so much of science is practiced with oral rather than written agreements. And the public, as well as the scientific community, rely on scientists to make statements supported by facts, statements that are not biased by political views or the financial repercussions for companies. A deliberate violation of this sacred trust is never forgiven in the scientific community.

While integrity is essential for all scientists and technical professionals, it can lead to some dangerous situations, particularly for those working in industry. Generally, the community of science expects its members to resolve differences in a polite and professional manner. If, for instance, a scientist uncovers a problem with a product that a company is about to release, we expect that scientist to bring it to the attention of those responsible in the company and not to issue a press release. Admittedly, there are rare occasions in which scientists have been forced to become whistle-blowers, but the community expects this to be the last resort, taken only after all other official avenues have been exhausted.

Scientists tend to value consistency more than many other people, so consistency is a good trait to cultivate. Some science professionals think about it in terms of defining a personal mission statement such as "my goal in life is to improve the quality of life for hearing-impaired persons." Consistency then requires using their mission statement as a touchstone for the appropriateness of any action. In other words, they regularly ask themselves "are my actions consistent with my personal mission?"

Another personal trait that helps people advance is a high energy level. While it clearly is more a matter of perception than reality, people who seem to have a low energy level are assumed to be accomplishing less and to care less about issues than their high-energy colleagues. Some people just display more energy than others, and it's difficult to imagine changing this particular trait. However, at least for important meetings or presentations, it can be quite advantageous to display a high degree of intensity and engagement, so as to be viewed as being passionate, committed, and ready to work.

Successful people never leave their future entirely in the hands of others. They proactively position themselves for advancement and guard their reputation. They are aggressive about themselves, in that they make sure people know what they are doing, what has worked well, and how much effort it took to accomplish. When opportunities arise, they seize them and use them to personal advantage. Although aggression sometimes carries a negative connotation, it can be appropriate and useful in a career, particularly if the extreme opposite is seen as avoiding conflict, missing opportunities, and approaching one's career passively.

Scientists generally are perceived to be less aggressive than most other professionals, but the advice that a certain level of aggression is valuable still holds true. The trick is to determine the appropriate level of aggression for your organization and your goals. This is certainly influenced by gender norms and the expectations that they bring with them. Thus, two people behaving in the same manner might be perceived very differently: the man simply as "aggressive" and the woman as "bitchy."

There are techniques that can be used to be appropriately aggressive while avoiding negative labels. For instance, it is completely appropriate for a scientist to tell her superior what sorts of opportunities she would welcome, for example, by stating in a conversation that it would be appreciated if she were sent to a workshop or to represent the group at some important meeting. In the context of a one-on-one conversation, this might well plant the seed for future decisions that won't require additional prodding. It is also possible, in most cases, to appear fair and magnanimous while getting what you'd like; it's simply a matter of how you phrase it. For example, the following two comments, which communicate the same desire to participate in a leadership workshop may inspire very different reactions when delivered at a faculty meeting: (1) "I'd like to go" or (2) "I have spoken with a colleague who found this interesting last year so I'd recommend we send one of us this year. I'm certainly happy to be among those considered for this workshop." While both express your interest, the latter choice is worded in a more expansive manner and will be seen as less aggressive.

Among the useful personal traits to cultivate is the ability to handle conflict in a positive, productive manner rather than becoming shrill or bitter. The thoughtful leader recognizes that the way to win on issues is to convince key leaders of the wisdom of your views and to hold the dissenters close to you rather than avoiding them. Particularly if you are in a position of leadership, it is far better to keep an eye on those who clearly disagree with your view and to deal with their disagreement than it is to let them voice it in a public manner that could undermine your authority. Not only is conflict inevitable in all jobs, it is the core of science, which relies on polite debate of the facts and the relevant conclusions from them. Reasonable people can rationally disagree on interpretations and facts; it never is wise to take such differences personally. Your colleagues will learn much about you from the way that you cope with conflict. Did you become defensive and shrill? Did you retreat and grow silent? Did you acknowledge the disagreement and thoughtfully seek a way to compromise or resolve the issue? Obviously, the last response is the one that marks you as someone who can be trusted to deal with disagreement and thus someone who might be a good candidate for promotion. In practice, a reasoned approach to conflict means never unleashing harsh words at a colleague. In fact, even when you are certain of an answer, it can be advantageous to ask for time to consider your response, simply to give the impression of thinking carefully about the issue.

Finally, there are many personal traits that have no relevance whatsoever to job performance, but that affect chances for advancement nonetheless. These are examined in the chapter "Personal Style" and include items such as style of dress, use of language, and level of conformity. Scientists generally have a well-earned reputation for being unstylish in their choice of clothing. This is particularly true in university settings where many purposely cultivate very casual and comfortable dress styles. However, one's choice of attire does send a message and, for women in particular, it is important to understand how this is perceived. Women in male-dominated fields often struggle to be taken seriously by their male colleagues. "Dressing down," even if it is part of the accepted norm of the group, can exacerbate the problem. At the opposite extreme, dressing too formally can be seen as an attempt to set oneself apart and can lead to a negative reaction. Thus, a good rule of thumb is to dress at the formal end of whatever is the accepted norm for your particular context.

The use of language by scientists has nothing to do with their technical ability but has a significant impact on how their colleagues and superiors perceive them. It is rare to meet scientists who are prone to using slang or colorful profanity. For women, in particular, to violate this rule is to risk being perceived as crude and outlandish. Additionally, because becoming a scientist requires many years of postsecondary education, the professional scientist typically takes pride in her writing and speaking ability (even when it is undeserved). It is important to be able to produce grammatically correct syntax to avoid the disdain of colleagues.

It is interesting to note that while we scientists consider ourselves unconventional, our very unconventionality has developed conventions. Particularly for women, who stand out simply because they are so underrepresented in the profession, being wildly unconventional can draw attention for the wrong reasons. It is worth thinking about the habits that you develop in this respect, with full appreciation of the ramifications. For instance, the gum-chewing, blue-haired female scientist rollerblading through the hallways will certainly stand out—but perhaps not in ways that will help her advancement. It is highly likely that she will be seen as too outlandish to be taken seriously as a scientist, even though gum chewing, hair color, and mode of transportation have nothing to do with technical abilities.

LEARNING FROM FAILURES

There are very few certainties in life, but something you can be sure of is that you will make an error at some point in your professional career. It may be something as minor as calling a person by the wrong name in public or as major as making a serious technical mistake in a published paper.

Errors have a great deal to teach us about the culture of our place of employment, about our colleagues, and about our own resilience. For instance, some work environments are cutthroat, with colleagues who actively try to undermine competitors; in others, this sort of behavior would be viewed as rude and inappropriate. The culture at your organization will become clear when you or someone near you makes a public error. If you are caught in an error, will your colleagues support you, remain silent, or publicly castigate you? The answer tells you about them and how much you can trust their integrity and loyalty. And how you handle the discovery of an error conveys much about you. Will you hide or openly admit the mistake? Your choice shows your colleagues the extent to which they can trust your integrity.

Failures are similar to errors in their effect on advancement, but they have a different sort of flavor. Errors refer to blunders or mistakes that should have been avoidable. Failures refer to attempts to address an issue without meeting with success. Sometimes failures are the result of poor planning or implementation by an individual. Sometimes they result from circumstances beyond our control, such as when a donor or funding agency fails to honor a commitment to support a project that subsequently folds due to lack of financial resources.

The way you handle errors and failures is important for job advancement because it reveals to your superiors how well you handle difficult situations and accept responsibility. Because the culture of science highly values honesty, it is never a good idea to deny an error or failure. Scientists are persistent researchers and they won't stop asking questions until they fully understand what happened. Rather than try to hedge or deny, it is usually better to admit quickly to errors and failures, apologize for the problems caused, suggest resolutions that minimize their ramifications, and then implement those resolution strategies.

You are likely to be forgiven the occasional minor error or project failure, but there certainly are exceptions to this rule. Financial improprieties, for instance, will likely cause significant trouble from which there is no easy way to recover. The general rule of thumb is that the scientific community puts great credence on your perceived motivations. Thus, a failure due to circumstances beyond your control (for instance, the canceling of a service mission to the Hubble Space Telescope causing the Hubble project to fail prematurely) is unlikely to cause much harm. Errors or failures resulting from neglect or incompetence might even be overlooked long term, although you may not be a prime candidate for high-level responsibilities for some time. However, problems that stem from illegal or unethical behavior, particularly if you are perceived to be seeking personal gain, will earn you sharp censure from the community of your peers and the aura of mistrust is likely to surround you for the remainder of your career in science.

What is most important about errors and failures is to learn from them. While forgiveness and understanding are common, you won't be forgiven for making the same error repeatedly or for having a track record of projects under your supervision failing. Thus, it is important when an error is uncovered or a project fails to spend time assessing what went wrong and how it might have been found earlier. This approach to your work is not unlike that taken in the morbidity and mortality conferences held in medicine in which scientists analyze clinical failures with the aim of preventing their recurrence. When the failure or error seems to be the result of a structural problem (such as not spending enough time in the laboratory checking the results of students), it is imperative to make changes in the way that you operate to avoid repeating the error.

There is life after a failure or error, even life as an active scientist or scientific administrator, unless the severity of the error and the sheer level of incompetence are very high. What the scientific community generally expects is an uncompromising focus on honesty and integrity. Finding an error in your work, for example, might involve retracting a paper under review or even writing a letter of explanation to be published about an article that already has appeared. Generally, the expectation is that scientists admit culpability, accept responsibility, and promise better in the future.

For those in leadership in science, it is important to understand the nature of the positions that they occupy. Best described as middle management, the work requires reporting to people whose responsibilities generally extend well beyond those related to science, as well as earning the respect of those in the unit in order to be effective. When things go wrong, they can do so with lightning speed and recovery can become impossible. This is particularly true if one has made enemies in the process of leading and making decisions. Many, many examples of this are reported in the business pages. Often a CEO is spoken of as though he walks on water—until he is found to have committed some blunder; then he is portrayed as never having been able to tie his shoelaces. The same sort of thing often happens in science, though on a somewhat less extreme level. However, setbacks are often temporary: A return to pure science from a management position or a move to a new place in order to continue being a leader is generally all it takes for your sins to be forgiven and forgotten.

It is perfectly normal to experience personal disappointments related to work at some point in your career. Imagine, for example, that you've finally agreed to become a candidate for a promotion after significant, persuasive efforts from a professional recruitment firm (i.e., headhunter). You do your homework, put your best foot forward, and have a great interview, but the job is awarded to someone else. This inevitably evokes disappointment, possibly coupled with embarrassment because your superiors might be aware of the situation. How you handle it will communicate a great deal about you, so it is important to develop a good attitude.

As with errors or failures, there are two important aspects to such matters: how you deal with them publicly and what you learn from them. Publicly, it is critical that you maintain your professionalism and normal demeanor, regardless of personal matters. If a particular situation is widely known, it can be wise to go public with your version of the story before rumors spread. Privately, it also is prudent to put disappointments in proper perspective, learn from them, and adjust your behavior (if necessary) in order to avoid the disappointment in the future. In this way, disappointments can teach us about our values and our personal styles and help us make adjustments appropriate to the workplace.

THE THEORY OF PUNCTUATED EQUILIBRIUM

There are many people who advance in their scientific careers by moving up steadily. By choice and hard work, by luck, or both, these individuals advance along a path that always is climbing, with few or no interludes where they seem to backtrack or move laterally. There also are many people for whom steady progression up the ladder comes at too high a personal cost, so they choose to move up more slowly and may, occasionally, even move laterally or backwards. Among this latter group, women with families are particularly prominent; with young children, they may well make choices predicated, in part, on compatibility with their family responsibilities or decide to forgo some opportunities entirely.

There are significant data that suggest that choosing to forgo career opportunities has a deleterious effect on one's career progress, at least in the short term. However, there also is an increasing body of literature that points to the growing acceptance of careers that progress by unusual routes and that are characterized as periods of great professional activity, punctuated by times of low activity. In evolutionary biology, there is a theory that applies well to many careers and should give people who have multifaceted lives cause for great optimism.

The theory of "punctuated equilibrium," which earned Stephen J. Gould great recognition, suggests that change is not uniform in pace but often occurs in cycles in which great transitions take place during short time frames and are followed by long periods with almost no change. As a model of career development, it suggests that it is not necessary to be a "superwoman" (i.e., a stellar researcher, world's greatest mom, devoted spouse, loving child, star athlete, super cook, and good neighbor, all at the same time). Rather, it suggests that it is acceptable—necessary even—to decide which of your many activities and responsibilities will assume primary importance at any one time. With the increasing awareness of the pressures of life, particularly on women with young children, there is hope indeed in the growing acceptance of careers that take the unusual pause or sideways move.

When balancing professional and personal responsibilities, it is important to preserve good will at work. Observing the following informal guidelines can help. First, if responsibilities outside the office and laboratory are going to affect the quality or quantity of your work, then you owe it to your colleagues to make it known—this is true for men as well as women and for short-term as well as long-term issues. There is a strong tendency in science to compartmentalize one's life and to hold personal matters as truly private, as things your colleagues may not know. But if personal issues (e.g., a sick child, failing parent, deteriorating marriage) are going to affect your work, then your superiors, at least, need to know. The best approach is to briefly describe the issues and their probable duration and anticipated impact on your professional responsibilities and to recommend actions that will work for you as well as your colleagues.

Second, if you find yourself in a situation in which you will be distracted from work for a significant period of time, it would be prudent to consider whether there are moves that would put you in positions less likely to negatively affect others in your organization. Stepping down from a project leadership position, for example, or moving from a committee responsibility for which you have few staff resources to one for which there are many (thereby giving you more flexibility with your time) are options to consider seriously. While moves of this sort are certainly a form of backtracking, they can be viewed as acting responsibly in the face of circumstances that require you to make difficult choices.

Finally, when discussing with superiors alternative arrangements for your work, always aim to offer constructive suggestions; these are always appreciated. If, for example, you find you need to step out of a position of responsibility, it is reasonable to offer the name of someone who might take on the role that you are vacating. Likewise, if a change in a structure would enable you to continue contributing, then suggesting this change, along with its compelling logic, is quite acceptable.

Scientists who opt for a different career path in response to the demands of children are not alone in the choices that they must make. Those who desire to stay linked to a spouse or domestic partner face similar tension between their professional and personal lives. Two-career families—once a rarity—are now the norm, and it is increasingly common for job transitions to require two new positions rather than one. This means that sometimes a move will be motivated not by a great opportunity for you, but by one for your significant other. While the dual move has become more common, it still is more difficult to arrange than a single career opportunity. However, handled carefully, the transition can work to the advantage of both people.

The most common form of dual career move involves one person who is greatly valued and a trailing partner who is held in somewhat less esteem. By making it clear early on that it is a "both or neither" situation, it is possible to put pressure on those recruiting the star of the pair to help find suitable options for the partner. This is true regardless of whether the pair of you seek positions in the same organization. Key in such moves is assessing what it will take to ensure that the trailing partner is afforded appropriate respect in his or her new position, particularly if you end up at the same organization. In science, this normally means giving thought to what the trailing partner will need to help establish his or her independence and what will increase the probability of success. All such issues should be on the table for negotiation. Further, while the negotiations for the two positions might go on simultaneously, it is absolutely critical that each person represent themselves.

Finally, one important piece of advice for couples who work in the same scientific areas: Although you may have met through your work and have been very successful in collaboration, it is extremely difficult to maintain a long-term professional collaboration without others making assumptions about who dominates scientifically. For the well-being of both members of the pair, it is advisable to have at least one project that does not include collaboration with your partner in order to establish your own reputation.

PRESERVING THE ABILITY TO MOVE BETWEEN THE TWO LADDERS

For much of one's career, it is possible to preserve the ability to move between the technical and managerial career ladders, and many scientists do so. For instance, many faculty members serve a term as a department chair or associate dean; others in industry work for a time as officers of their professional societies. Moving back and forth is wonderful because you can try out different kinds of skills without being forever removed from the laboratory.

The ability to make such shifts is greatest at the lowest levels of the career ladders, with the gap between them growing quite wide at the ranks of upper management. What is interesting is that the resistance to moves from one path to the other is not the same in both directions. It is far easier to move from a purely technical position into one that involves some management than to go the other direction. This largely comes down to the recognition that the skills required for science management are not the same as those needed to be a pure scientist. Scientific research requires one to have very detailed knowledge of a particular area and keep up with advances in that field. To be a manager, on the other hand, one must learn how to negotiate with people and motivate them. Time invested in this activity is time that one cannot spend keeping up with a technical area and making scientific progress. Thus, as one progressively assumes managerial duties, the time remaining for scientific accomplishments dwindles and one's attractiveness for a purely scientific position fades. When viewed from the other direction, however, the situation is different: Since managers spend most of their time dealing with people, they must have the respect of those who report to them, so being an excellent scientist is not a hindrance but an advantage.

Just as there are hierarchies in science organizations, there also are hierarchies in management and one is normally expected to advance through the system, one step at a time. Occasionally, exceptions are made for unusually gifted people who might discover an interest in leadership after many years of avoiding managerial roles. Nannerl Keohane, former President of Duke University and Wellsley College, for example, never served as an academic administrator at a level higher than a department chair before assuming the Wellsley presidency. Such exceptions are rare, and women tend to be overrepresented in them, partly due to the demands of child raising (leading to deferral of administrative responsibilities) and partly due to subtle forms of gender bias that can make it difficult for women to earn notice at early stages of their careers.

While it is useful to have the option of jumping between the purely scientific and the administrative career ladders, at some point you will be forced to choose between them if you wish to climb any further. Some real soul-searching will be needed because the decision will be much more difficult to reverse than previous decisions. But by this time, you'll usually have had sufficient experience of each to determine which holds the greatest attraction.

HELPING OTHERS HAS ITS OWN REWARDS

The topic of mentoring was addressed earlier in this chapter and is the subject of another chapter in the book. However, the flip side of mentoring—serving as a mentor—also deserves some attention. It is absolutely true that people need mentors at all stages of their careers.

One reason to serve as a mentor is to pay back the debt that we owe to the mentors who have helped us; it is a kind of passing of the baton. Mentoring relationships tend to be warm and friendly, so at minimum, we have an opportunity to introduce good relationships into our professional lives. And needless to say, the sheer joy of helping another succeed can be rich succor. Another reason to serve as a mentor is because it is quite possible to learn from those more junior to us in experience and rank. We are never too old nor too wise to learn from others (although some of us may be too arrogant). Also, we can build up a network and a cadre of supporters through these relationships, a group of people who can advocate on our behalf at later stages in our careers. Not everyone advances at the same rate. By recognizing a rising star early on and helping him or her to succeed, you might find yourself with a boss who feels that he or she owes a significant part of their success to you. Ideally, the best situation is to find yourself on some rung of the career ladder, surrounded above and below by people whom you've helped at various stages in their careers.

Each of us has a limited time during which we work as productive scientists or scientific managers. Through mentoring, we extend our impact by passing on our values, methods, and insights to those who follow. Mentoring also can play a key role in "succession planning," the process of preparing your organization for continuity when you step down from your current position. In some organizations (particularly businesses), and generally at the higher levels of career advancement, succession planning is expected and valued. At academic institutions, it is unheard of as a formal process and often neglected even at informal levels (which can cause great problems). The important point to remember is that, by planning for your successor, you will be helping your organization to continue moving forward on the path that you laid out. That alone is a significant reward.

CHRISTINE S. GRANT, Ph.D.
Professor, Department of Chemical Engineering
North Carolina State University, UNITED STATES
Chemical Engineer

Dr. Grant is an accomplished researcher in the area of surface and interfacial phenomena as they relate to industrially relevant problems in green chemistry and chemical engineering. A graduate of the chemical engineering programs at Brown University (Sc.B.) and the Georgia Institute of Technology (M.S. and Ph.D.), she is one of fewer than 10 African-American female chemical engineering faculty in the United States. Dr. Grant is a Boeing Senior Fellow of the Center for the Advancement of Scholarship in Engineering Education (CASEE) and has served as a member of both the Board of Directors and the Chemical Technology Operating Council of the American Institute of Chemical Engineers (AIChE), the premier organization for aspiring and practicing chemical engineers.

Her long-standing commitment to excellence in science and engineering is demonstrated by her considerable leadership involvement at the local, state, national, and international levels. At North Carolina State University, for example, she has served as advisor to the student chapter of the National Organization of Black Chemists and Chemical Engineers, and Director of a five-year "Research Experience for Undergraduates" program (funded by the National Science Foundation). She also is the driving force behind an initiative to develop a focus on the successful recruitment, retention, and promotion of qualified engineers and scientists from underrepresented groups to faculty positions. A former Chair of the AIChE Minority Affairs Committee, she continues to act as advisor to many national committees and study groups regarding the perspectives of women and minorities in science, mathematics, and engineering technology.

Professor Grant has established an extraordinarily outstanding record of achievement in her mentoring activities at home and around the world. She was recognized for her advocacy for women in science and her educational outreach to Ghana, West Africa, by induction into the YWCA Academy of Women Award in Science and Technology, and Sigma Iota Rho, an International Honor Society. In recognition of her singular and sustained commitment to mentoring at all levels in the academy, she was honored with the 2003 National Science Foundation Presidential Award for Excellence in Science, Math, and Engineering Mentoring.

C h a p t e r

5

· · · · · · · · ·

MENTORING

· · · · · · · · · · · · · · · · ·

Christine S. Grant

SCIENCE: A MENTORING PROFESSION

Science is one of the most exciting, rewarding careers that one can have. The quest for information, knowledge, and the solution to both easy and complex problems is what scientists and engineers do best. While there are differences in the way scientists and engineers view the world, as the various fields become more complex and research becomes more interdisciplinary, these differences are diminishing. For example, I often describe a chemist as one who works on the interactions of molecules at the atomic level and the synthesis of new compounds and who becomes excited at the development of a new material and a chemical engineer as one who works on the processing of the chemists' discoveries on a large scale. The traditional chemical engineer thinks about the heat transfer, fluid flow, and reaction kinetics in 3000-liter reactors and works with chemists to facilitate the economical, environmental, and safe production of these materials. But wait—there is more: Now chemical engineers are working on the nanoscale chemistry of electronic materials, the biochemical engineering of new pharmaceuticals, and environmentally conscious manufacturing.

What does this have to do with mentoring? Having access to current, detailed information on cutting-edge technologies and emerging areas of

85

research is critical to making effective career decisions, but if you rely on published information only—however excellent it may be—your knowledge of your options will be incomplete. Who better to inform you of the potential directions of science and engineering but people who are actively involved in the work? Who better to offer perspective and advice on the possibilities for your career in science than those who already have first-hand experience? And though it may be more helpful or comfortable at times to have mentors who are like you, this matters less than what they *know*.

The very nature of science requires mentorship. Scientists are often portrayed as isolated individuals who lack personality and social skills, but we all know this is not accurate. If one thinks about how a biology laboratory operates, how a chemical plant is run, or how a nanotechnology think tank works, for example, one realizes that all require considerable interaction with colleagues. Often there is a senior scientist, responsible for the overall intellectual productivity of a group of individuals. There may also be a management team associated with the implementation and/or commercialization of an idea, process, or product. At every step, information is exchanged; knowledge is transferred. It is critical for scientists at every stage of their careers—but particularly as they begin—to understand the dynamics associated with the *process* of doing science in their particular fields and contexts, as well as the associated *political issues*. Mentoring greatly facilitates this.

> As a promising engineering student, the Regional Chairperson of the student chapter of our professional society was someone my colleagues and I really looked up to. From her, we learned how to run the organization effectively. Even when she became the national chairperson of the society, her example was inspiring. She handled herself with style and poise in difficult situations. Eventually, we ended up in graduate school together. Though she was a year ahead of us and in a different department, we still thought of her as our mentor. She helped us make a smooth transition to the highly competitive atmosphere of this research-intensive university and advised and supported us through our doctoral qualifiers, selection of research advisors, and eventually the completion of our doctoral theses. Somewhere along the way, our reverence and awe grew into admiration and friendship.

In this chapter, we introduce several types of mentors and mentoring relationships, describe a reasonable set of expectations for mentors and mentees, and offer strategies for identifying potential mentors, for initiating and maintaining the relationships, and (when they have reached maturation) for making the transition to colleague status. These are as relevant to students and early career scientists as they are to well-established professionals; mentoring is beneficial to all. While some mentoring can occur spontaneously, we hope that you will take an active role in your own success by initiating and maintaining effective, healthy mentoring relationships.

WHAT IS MENTORING?

Mentoring is a term used to describe the positive interactions that occur between an experienced and trusted advisor (mentor) and a less experienced individual (variously termed mentee, protégé) or group of individuals that facilitate the professional and sometimes personal development of the junior person(s). It empowers the mentee to move forward with confidence along his or her career path by providing him or her with support, encouragement, insight, advice, and (often crucial) information for making informed decisions and the (often difficult) choices that will lead to professional success and satisfaction. Mentors are some of your strongest advocates.

Mentoring enhances your ability to be effective and productive in your current and future endeavors by helping you to identify the skills, knowledge, and experience that you need. It eases your transitions (see the chapter "Transitions" for definition and discussion) by helping you to understand the "hidden" rules and conventions of work. Your academic training will have provided you with a solid foundation upon which to build your career, but how to use it most effectively in a given context requires information that you do not have. A well-chosen mentor will know. Learning occurs differently for all of us and usually involves a combination of hands-on experience, reading and reflection (e.g., of/on the literature in our field), observation of others, and direct instruction. Mentoring is an efficient way to learn about the nuances of the systems in which we work.

Mentoring is *not* the unhealthy control or domination of one person (or group) by someone with greater experience and power. It is not the manipulation of a subordinate for professional gain, nor is it the abuse of a connection with a person in a prominent position. Rather, it a relationship built on mutual trust that unites people with common interests (regardless of their backgrounds, differences in status and power, etc.) and benefits both. While the advantages for mentees are obvious, mentors enjoy different rewards, including the satisfaction of seeing a protégé succeed and the sense of accomplishment that derives from that.

TYPES OF MENTORS

Mentors can be grouped into three broad categories, according to the type of information and support they offer. As will soon become apparent, it is quite appropriate to have more than one mentor at any one time and a series of mentors over a lifetime.

Peer Mentor

A peer mentor is a person who will help you learn the ins and outs of your current position: someone who has the same work responsibilities as you but who has been with the organization a bit longer or someone one level above yours. They may have performed your duties before being promoted and would benefit directly from you fitting into the system quickly and performing well. Suppose, for example, you are an engineer who is part of a team working on a major project. Your mentor could be the lead engineer on the project; since her role is to guide the project to successful completion in a timely fashion, she would be highly motivated to mentor you through the early stages of your transition to the job and the team.

When Sally started her position at the Geological Survey in Colorado, she was really excited about the opportunity to make a difference in the ecological landscape of the local region where pollution was a major problem. At a meeting of the young geologists' society, she was pleasantly surprised to learn that there was another person in the Society who graduated from the same university. While they were assigned to different remote posts in the forest, there was a similarity in their job tasks, so she decided to talk with him about his work. Tom, who had joined the company about eight months earlier, initially seemed a bit hesitant to share the details of how he did his job. She wondered if he may have felt threatened by her forward approach and probing questions, so she decided to meet him for coffee at his site—under the pretext of seeing a particular piece of equipment that she needed to buy for her own survey location (which she needed to do anyway).

During their conversation, she openly told him that she hoped they would work together to strengthen the regional geological activities and that she wanted to learn from his experiences in the job, so that she could be equally effective as an employee and colleague. After some discussion, he agreed to meet with her once a month for the next year to exchange information on the activities at their respective locations. By the end of the year, both locations had improved the rate at which they were meeting the pollution reduction goals for the region. Sally made sure that Tom's boss knew that she appreciated Tom's input and their collaborative working relationship. Though she could have obtained similar training and expert advice at her own site, she realized that having a peer mentor at a different location was beneficial to them both. There was less possibility of interpersonal conflict and competition because both were able to excel at the same tasks in their respective regions.

Career Development Mentor

Having access to someone who has an intimate knowledge of the system in which you work and the expectations of employees at various levels in the organization is invaluable when making important decisions about your continued growth and upward mobility (if that is your goal) and is crucial to managing your career. Usually, a career development mentor is someone much more senior in your organization, often a person in management. He or she also can be someone in the same industry, but outside your organization, who has similar knowledge and expertise. You may meet on a periodic or regular basis (annually, for example) to review your curriculum vitae and discuss the steps required for you to position yourself for the next career move. Mentoring could also take the form of an ongoing dialogue on a weekly basis through which you could learn, e.g., how to successfully negotiate more laboratory space, new project assignments, support for additional training (e.g., an advanced degree), raises, and promotions.

Early on in her tenure-track position at a research-intensive university in England, a young biochemist met with the head of the department to discuss her research goals and career plans. The head had no doubt that she had the potential for success in all three areas of responsibility: research, teaching, and service. But when she realized how enthusiastic the young woman was about contributing to the institution and to the larger science community, she foresaw a great potential for her to become overcommitted to activities that would compromise her ability to establish herself in her field and earn tenure. She saw too many new academics—especially women—burn themselves out, and she did not want this to happen.

So, she advised her to focus on building her research program, attracting funding, and publishing. As was standard practice in that department for all new faculty, she assigned her a lower than average teaching load for the first three years of her appointment and suggested what a reasonable number of committee responsibilities would be. Though the younger scientist knew it would be difficult to curb her desire to become involved in activities that were important to her (but peripheral to her primary goal for the first few years at the institution), she followed the advice.

Thanks to the support and guidance provided by this mentor and her own discipline and focused efforts, the young biochemist succeeded in establishing herself in her field and earned tenure ahead of the usual time frame. The choices she made at the outset of her career protected her from overcommitting herself and prevented her from setting the dangerous precedent of saying "yes" to every invitation from the scientific community. Now in the middle of her career, she provides similar career advice and mentoring support to her own graduate students and early career scientists.

Personal Mentor

A personal mentor is one who offers support and input on issues that are not directly related to science, including the critical issues surrounding the delicate balance of work, family, social, and civic obligations. If, for example, you are hoping to take a sabbatical leave from your job to enhance your educational portfolio, you could talk to a person who has successfully negotiated time off to earn an advanced degree. If you are suddenly facing the management of the health care of an aging parent, the experience, advice, and support of colleagues with young children who have negotiated reduced work responsibilities or alternate work arrangements (e.g., shifted work hours or telecommuting) may be encouraging and helpful.

Even before she joined a national, governmental laboratory in North America to head an important research group, a young, ambitious geologist knew that, one day, she would like to have a family. Her background investigation of the organization lead her to believe that this particular environment would provide the intellectual challenges that she craved and the physical resources that she needed to pursue her professional goals. It also reassured her that well-established policies and procedures were in place for employees wishing to take parental leave. It seemed a perfect fit.

Before accepting the position, she sought the input of her career mentor, for his insight and perspective had always helped her to clarify her thinking, priorities, and goals. His advice influenced her decision to spend her first years in the organization establishing herself in her new position, building her laboratory and research program, and earning the respect of her colleagues within the system and beyond. Another suggestion that he offered proved to be most helpful in her decisions about family: He strongly encouraged her to develop a personal rapport with the other women in the organization, particularly those who had families, for, he reasoned, only they knew the realities of working in that environment and raising children—written policies and procedures are one thing; how people are perceived and treated if they take advantage of parental leave is quite another.

She followed his advice and gained much insight. One senior scientist with whom she spoke began her family when she was completing her Ph.D.—even before joining the organization—and did not take any time off from work. Another decided to leave her job in industry to raise her children full time and, after 10 years, upgraded her bench skills and returned to full-time research in this government agency. A mid-career scientist and her husband (who works in academe) decided after starting a family that one person's career would take precedence for a time, and then they would switch. While her husband worked toward earning tenure, she maintained reasonable progress in the laboratory but did not push for advancement in her government job, so she could play a dominant role in child rearing. After her husband earned tenure, he negotiated a reduced teaching load (so he could maintain his research profile) and assumed more responsibility at home, while she intensified her research efforts and worked toward a promotion. Still another colleague, newer to the system, has taken a year's maternity leave and is maintaining close contact with her laboratory by e-mail and regular visits.

By establishing a rapport with her female colleagues, the young geologist learned about the realities of balancing professional and personal life in this context and was able to benefit from their advice as personal mentors: Regardless of when she begins her family or how she balances the myriad responsibilities, if she wishes to advance her career while raising her children, she needs (1) a strong, personal support network (e.g., spouse, extended family, friends) and (2) (more importantly from the perspective of her employer) to continue to make progress in her research. By being proactive, she had already developed her own support network within the system—and even identified people who have turned out to be excellent peer mentors, career mentors, and personal mentors.

One of the challenges with a personal mentor is to maintain the professional decorum expected in your job environment. The best results come from being clear about the boundaries of the relationship and continuously reviewing the potential ethical implications of the behavior. This is not to say that long-term personal relationships cannot develop from what began as mentoring relationships (e.g., deep friendships, marriage), just that there are business "rules" and social conventions that must be respected.

A Mentor for All Seasons?

Obviously, one mentor cannot be all things to you. Your needs will change throughout your life. You may have one mentor for career coaching in your particular company and another one for the pursuit of your graduate degree outside of the work environment. You may find yet another for the technical aspects of your work. It would be helpful if at least one of your mentors were able to identify with you through what I will call "a shared experience." For example, suppose you are experiencing a difficult pregnancy and your supervisor gives you more work each day leading up to your four-month maternity leave. You know that the demands that he is placing on you would be unreasonable for anyone and you need to address the issue. But you do not know how best to handle it; the last thing you want is to give the impression that you are using your pregnancy as an excuse. Then you remember a conversation you had with a fellow scientist from another pharmaceutical company who attends the same aerobics classes: she experienced a similar situation five years earlier. She might be willing to talk you through the particulars of your situation and suggest specific strategies for dealing with your boss in a professional and effective manner. She may continue to mentor you as you go through the newness of motherhood and eventually the difficulty of seeing the first child off to elementary school. It is your shared experience of pregnancy and motherhood as working scientists that could inspire the relationship and make her a most effective mentor.

STRATEGIES FOR FINDING A MENTOR

A mentor can be outside of your immediate professional or academic domains or be directly associated with your current environment and able to influence people on your behalf as you move forward in your career. All will provide support, encouragement, and advice. Regardless of who you are or at what stage in your career, having access to effective mentors can enhance your success and satisfaction. In this section, we suggest criteria for identifying who might be the best mentors for your needs and discuss ways in which you might meet them.

Identifying a Potential "Match"

The first step in this process is to identify a prospective mentor. There are several important issues to consider: your personality, circumstances, and needs; the availability, accessibility, and willingness of the potential mentor to assist you; and, more importantly, how well the two of you "match." Obviously, they must have expertise in the area(s) for which you are seeking advice and support. And there needs to be compatibility between you, in terms of both personalities and communication style. Mentoring relationships are built on trust; hostile relationships are detrimental to both parties. As mentioned earlier, having shared experiences is helpful in mentoring relationships (though not critical).

Obviously, if potential mentors have been helpful to others, they may be good candidates to mentor you. Make inquiries of their past and current mentees. Are they comfortable with their mentor's style? Is their mentor truly interested in the development of the mentee? Are they available or so overextended that the mentee rarely has access to them? (This does not mean to imply that mentors should be at your beck and call, but a growing trend of inaccessibility during critical moments in your professional life is not a sign that this person would be a good "fit.") Are they able to "let go" of the relationship when appropriate? The answers to these questions may help you to identify who might be effective candidates for mentoring.

A Mentor Who Is Just Like Me?

It is important to understand that your mentor does *not* have to be just like you for the relationship to work well. The person may be male or female, young or old, of the same or a different race, culture, religion, temperament; he or she may be at the same company, institution, or agency as you, or work in a different field altogether. The important point is to find someone compatible who is sensitive to your unique job or educational situation, has the appropriate knowledge and experience that you seek, and is interested, willing, and able to participate in a mentoring relationship with you.

Everyone said she would be the "perfect mentor". There may be a person who appears to be the perfect fit for you: She may have graduated from the same university, played lacrosse (just like you do), and even aspired to (and now has) the same high-level management position that you are aiming for in your career. People may even tell you, "Oh, you have so much in common with Jane; I suggest you ask her to mentor you. She is really approachable and has been through some of the same situations that you are experiencing now." Though it sounds as though she would be a perfect candidate to mentor you, make sure you investigate fully; you may be surprised. It may turn out that Jane's professional and personal responsibilities are such that she has no time to mentor anyone. She may be under tremendous pressure to finish a critical project at work and cannot afford to invest herself in activities that are not central to her own career development. Jane may not have had much opportunity to be mentored when she was becoming established and believes that you should be able to do the same. Or she may see you as direct competition for the job she currently holds. It may also be that, in spite of the perceptions that others have of you and Jane, your personalities simply may not be compatible.

"How Could He Possibly Be a Good Mentor for Me?"

There are differing views on the "usefulness" of mentors who are not like you. Some argue, for example, that a man could not possibly mentor a woman: "He has no idea what women go through in their professional and personal lives." This perspective implies that without first-hand experience of gender bias (for example), men can have nothing useful to offer in a discussion about how a female mentee can deal with these situations herself. Others could argue that significant differences in ages would present barriers to communication because of "the generation gap," or that someone from a large urban community in the northern part of the country could not possibly assist another from a southern rural town. Granted, there is some truth to the underlying premise: our experiences in life *do* help to shape who we are and influence the expectations and (preconceived) notions that we bring to the workplace. We need to acknowledge this—in ourselves as well as in others—when we approach any mentoring relationship, even (as we've seen) with people who seem the "perfect fit." After all, what is the purpose of mentors? Are they to be best friends who sympathize with all that we say because they identify with our situation? Or are they knowledgeable, empathetic supporters and advisors who encourage us to develop the skills that we need to deal with the situations that we are facing?

> Early in my career as a new faculty member, I was mentored by a senior white man who was a faculty member at another institution. As the relationship began, I discovered that I had an issue to overcome before the mentoring relationship could work. As an African-American woman, I had to acknowledge the fact that this person really did have my best interests at heart: he simply wanted to help me understand the nuances of both the technical and procedural aspects of being a faculty member so that I could achieve my goals.
>
> We first met at a national conference of our professional organization when I approached him about an article that I was reviewing for the journal that he edited. Eventually, we met annually at the national conference, to review my curriculum vitae and discuss my career goals for the next year. It was a great privilege to have such a busy and much sought-after person spend at least an hour with me each year and to receive excellent critical input into my applications for promotion and, eventually, tenure.

Meeting Potential Mentors

Having identified the kinds of mentors who would most closely match your current mentoring needs, you need to find the people. Mentors and mentees can meet in any number of ways, both formally and informally. While some mentoring relationships begin spontaneously, due to close proximity in working relationships or natural interactions due to shared interests or preexisting friendships, others need to be consciously cultivated.

"Naturally Occurring" Mentors

Oftentimes, a mentor meets an individual whom they feel compelled to mentor. The mentee may have distinguished themselves scientifically, thereby catching the attention of the mentor, who then becomes motivated to assist them in their career development. Or the mentor may see in the junior person an "earlier version" of him- or herself and may want to ensure that the mentee does not make the same mistakes that he or she did. Regardless, it is the mentor who offers advice and support, even before any "mentoring relationship" is established. The mentee may even be unaware that they are being mentored. In the following story, a microbiologist recalls the role that an (at the time) unknown mentor played in her being awarded a competitive national fellowship to attend graduate school. It demonstrates how an innocent meeting can result in a person finding a mentor for life.

> When I was an undergraduate, I had occasion to interview a first-year graduate student for the school technical paper. Soon after the interview, our paths crossed in the hallway of the biology department and she asked me how things were going. I did not think much of the encounter, I just mumbled some incomprehensible comment about the length of my last organic chemistry lab. She smiled, encouraged me to "hang in there," and quickly rushed off to her next class.
>
> After a few months of occasional and similarly brief meetings, she stopped me outside her office: "Say, Jackie, are you planning to go to graduate school?" I told her that I'd had a great job interview with a top pharmaceutical company in California and that I was prepared to accept the position, if offered. "You should consider applying for the National Biology Fellowship Program. They pay your tuition and fees and provide a stipend for your living expenses." At that point, I was still not interested but felt compelled to follow her to her office to look at the materials that she offered to show me. After describing in glowing terms the benefits of graduate school, she started to fill out the application form for me. "It never hurts to keep your options open," she said. "If you really are interested in staying in the area, you may want to consider looking at some graduate programs." I was taken aback by her actions. But here was someone who was so passionate about this opportunity for me, so certain I would succeed...how could I say "no"?
>
> I was awarded the fellowship, turned down the position at the company, and ended up earning my doctorate in the field as well. I am convinced that I would have never gone to graduate school without that chance meeting in the hallway years before and the dedication of a person who saw promise in me and a potential career match for my future.
>
> Over the past 20 years, our relationship has evolved into something that is mutually beneficial and we have become friends and colleagues in the field that is not known for having a large number of women in it.

As you see from this example, there *are* people who are interested in your success, people who *are* willing to offer advice, support, and encouragement. At first, you may be oblivious to the fact that a relationship is being developed (as was the case in the example above, where the mentor showed a genuine interest in the speaker and asked her about plans for the future). Sometimes you need to stop and ask yourself who is taking an interest in you. Who is asking questions about your current work and future plans? Who is showing concern about what you are doing and how well you are succeeding? Those are people who may actually end up being your best mentors.

You may even find that your supervisor, manager, or boss becomes your greatest advocate. While extremely helpful to your career advancement, having a mentor who is your direct supervisor may occasionally engender reactions in others (e.g., jealousy) that require mental toughness to deal with professionally (e.g., in handling unfounded rumors or misconceptions that may arise).

In my case, the mentoring relationship evolved out of a more formal boss–employee relationship. I admired his management style and scientific skills in the company and sought him out to learn more about the organization. I came to be under his direct supervision and as I began working my way up the ranks, I continued to seek his advice at critical junctures. During the entire time that I was under his supervision, I did not abuse the mentoring relationship and was very cognizant of the possible concern some colleagues may have had about favoritism.

Eventually I interviewed for, and was promoted to, his position after he retired. Though many of my colleagues were supportive, I was aware that there were a few who felt that I had been selectively groomed for the position and had been promoted because of privilege. But in reality, I had earned the position though my own performance, qualifications, and merit. My boss was not directly responsible for the promotion; it was the decision of a committee of upper level managers. I chose not to go out of my way to speak against the rumors and misconceptions. Instead, if people came to speak with me, I calmly explained the promotion process to them, and let my performance in my new job prove the appropriateness of the choice. A few years into the position, the naysayers were approaching upper management, singing my praises.

Apparently, my boss (and mentor) was aware that there had been significant opposition to my appointment from a few people who would work for me, but he did not tell me at the time because he knew it was groundless and could distract me from my course. It wasn't until much later that I learned just how much of a support mentors can be—whether you know it or not. They may act "behind the scenes" to protect you and speak up against or clarify misconceptions that do not accurately reflect your abilities, intelligence, or integrity.

Formal Mentoring Programs

Your university, company, or professional association may have formal, one-on-one mentoring programs in place to facilitate the mentoring process. You would be assigned a person (e.g., a faculty advisor, supervisor, or more senior work colleague) to be your career mentor, based on a similarity of research focus (or major), career goals, background, job function, and so on.

"Cluster mentoring" is another type of formal face-to-face mentoring program, though in this scheme, you become part of a group that is mentored by one person. The self-selection or assignment to the group is based on some commonality, such as academic background, career goal, work environment, special interest, or area of concern, such as gender, race, and age. In this model, a senior mentor or advisor may be assigned to each group, or a person who is only slightly past the level of the mentees, depending on the group.

Excellent online initiatives exist to help match mentors and mentees in the areas of science and technology, such as *MentorNet* (www. mentornet.net), an e-mentoring network for women in engineering and science, the first and largest of its kind, and *Just Garcia Hill* (jgh.hunter. cuny.edu), a virtual community for minorities in science. They may be sponsored by academic institutions, science foundations, granting agencies, professional associations, special interest groups and associations (e.g., for women and minorities in science), businesses, and the like.

As with every mentoring opportunity that you consider, do your research carefully before committing yourself or sharing your personal information. Make sure you are confident that the program is reputable and will fill the mentoring needs you have identified. There are some consulting programs for mentoring that charge a fee, but there are enough people and professional groups offering mentoring support that you should never have to pay for the service.

Meeting Potential Mentors at Scientific Conferences

If formal mentoring programs are not available to you, you will need to take the *initiative* (a wise strategy to follow in any event, to cultivate these relationships). Regional and national conferences of your professional organization are excellent places to meet mentors in your field. If you are a chemist, for example, the annual meeting of the American Chemical Society (an organization of over 100,000 members) attracts thousands of participants who include (1) students who want to learn more about the opportunities and jobs in the field or who are actively seeking professional positions; (2) new graduates and young professionals making the transition to their positions with a company, research institute, government laboratory, or university; (3) mid-career scientists who are headed for management positions; (4) senior personnel who are coordinating or managing large groups of people or strategic initiatives in an organization; and perhaps (5) an eminent scientific scholar in your field whom you have revered for years and often cite in your work. All who attend these conferences actively *seek* interactions with their colleagues and *expect* to be approached by people interested in their work. While they may not have the same agenda for their conversations, they certainly will be open to discussions about cutting-edge scientific research, advancing the profession, and issues central to career development.

The opportunities for actually meeting and speaking with potential mentors are many—and as varied as the conference program, for example, during (1) the question period following his or her seminar, (2) the poster session (either you or he or she could be presenting), (3) organizational business meetings, (4) pre- or postconference workshops or short courses, (5) special interest group meetings (e.g., in your subfield, women in science), and (6) informal social events including receptions, dinners, organized tours, and recreational activities. Since this may be the first time the person has met you—and first impressions can be lasting ones (as you will read in the chapter "Personal Style")—prepare carefully, plan and practice your approach, then follow through.

Identify Specific People. Well before the meeting, read the conference program thoroughly (often available online). Examine the list of keynote and plenary speakers, seminar and poster presenters, workshop and short-course instructors, and the like, for people whose interests overlap your own. (If the program is not available in full, read the list of symposia topics.) From this information, you will know who will be attending (or who is likely to attend). Members of your immediate network (e.g., supervisor, work colleague) may be able to suggest names of people from this list who may be good candidates for mentors. Referrals such as these may make it easier for you to approach the people because you can include this information in your introduction (see below). Once you have identified the people you would like to approach, recheck the meeting schedule when you arrive at the conference (changes to the program often occur) to confirm where they will be (and when), so you can be there.

Introduce yourself. Before the event, you may wish to contact the potential mentor by e-mail, to introduce yourself, briefly describe your interest in their work, and ask whether you can meet during the conference. By doing so, some of the awkward preliminaries will be dealt with, even before you meet face to face: They will know who you are and the interest you have in them and will be looking for you. At the same time, you will be reassured of their willingness to speak with you (because they agreed to a meeting) and can be confident that the time and location are appropriate and convenient (because they suggested it).

The first meeting may still be awkward for you (it is for many people, regardless of age or experience), but if you have prepared a few questions beforehand and stay focused on your interest in *them,* you will soon forget yourself. People always enjoy talking about their work (it helps them to overcome any shyness that *they* may have) and your genuine interest and insightful questions will demonstrate that you have done your homework. For example, *"I see you have been actively involved in research field <ABC>. Can you tell me about this particular aspect of your research <fill in the details>?"* Another question could be

"I would like to learn more about Society <XYZ>; this is only my first meeting since I joined my present company; what benefits have you enjoyed as a member?

At an appropriate point in the conversation, briefly tell them about your own research interests and career aspirations[1] (if you did not contact them before the meeting)—or reiterate what you said in your introductory e-mail—then ask a few direct questions to solicit their advice. For example, once they have answered the first question above, follow up with *"I have been trying to identify a person in my area <a subdiscipline of ABC> to be a career coach and mentor. Do you know of anyone who might be willing to fulfill this role?"* Or, to the second question you could add *"I understand that you are the coordinator for topical area <ABC>. I am interested in chairing a session in that area next year. Could you tell me how best to prepare my proposal?"*

See how they respond to your questions. Do not speak only about yourself but include topics that you have in common; you are, after all, hoping to establish a relationship with them. Look for cues that indicate that they are interested in what you are saying and are willing to offer support. Though you may wish to make a connection with them, do not expect them to be similarly motivated. Remember that they, too, have their own agendas for the conference. Be aware of this and do *not* interpret a hasty reply as a brush off or take such a response personally (the chapter "Mental Toughness" suggests strategies).

Follow-up. After this first meeting, you may initiate subsequent interactions (e.g., short updates by e-mail, casual conversations at other meetings) that convince you that this person may be willing and able to be a mentor to you. At this point, feel free to ask them directly. For example, *I have enjoyed our conversations to date and have appreciated the support and advice you have provided. Would you be interested in mentoring me on a more regular or formal basis?*

I have been following your work in the area of <ABC>. As you know, I am just starting out in my career and have been looking for someone to periodically provide me with honest feedback on the next steps. Do you routinely do that for people? Would you consider doing that for me?

I understand you are very busy in your work as Dean at <name the university>. [This demonstrates that you understand the value of their time.] Could we meet once a year—at this annual meeting, for

---

[1] It is a good idea to develop a 30-second "introduction" to yourself that you can use as an "opener" in any conversation. Make sure it is well rehearsed and sounds natural. (See next chapter for details)

example—to review my progress? I would be happy to send you a brief
summary beforehand, so we could keep the meeting short.

The approaches that you would take to initiate and maintain conversations with potential mentors are similar to those for any networking effort (see the chapter "Networking").

> When the new executive of the American Physical Society was announced on the home page of the organization, I learned that it included a nanotechnologist I had always wanted to meet. Since I knew that he would be involved in leadership meetings at the next annual conference, I planned my strategy accordingly. I found out from the conference program when and where the session would be held, then waited outside the room just before it ended to catch him as he left. While it was obvious that he was in a hurry to go to another meeting, I was able to introduce myself, give him my business card, and arrange for us to meet at a more convenient time. This led to a more successful meeting and began a long-term professional relationship.

MANAGING AND MAINTAINING THE RELATIONSHIP

A critical aspect of mentoring is the management and maintenance of the relationship, after it has been established. Fostering healthy professional and personal *networks*—or any relationship, for that matter—requires time and commitment on the part of both mentor and mentee. It involves mutual respect, sensitivity to each other's professional and personal boundaries, a clear understanding of the expectations and responsibilities of each party, open and honest communication, and some plan for your interactions. How this is achieved will be determined by the individuals themselves; there is no magic path to a successful mentoring experience. Most important to remember is to conduct yourself with professionalism and integrity.

Clarify Expectations and Responsibilities

Formal mentoring programs often have written guidelines for mentors and mentees, so that the terms and conditions of the relationship are clear (Table 1). This minimizes misunderstandings and the potential for abuse (by either party). In other mentoring relationships, such a structure may not be as explicit. If your mentor does not take the initiative, make it your responsibility to clarify your understanding of the terms and conditions of the relationship and your responsibilities and expectations and ask your mentor to do the same.

Table 5.1 Expectations and responsibilities of mentees and mentors

	Mentees	Mentors
Communication	• Ascertain the method and frequency of communication preferred by your mentor (e.g., phone, e-mail, meetings) and respect it. • Respect the position of your mentor (e.g., his or her title, "office," staff).	• Set boundaries for communication and be consistent (e.g., honor meetings, return messages in a timely fashion). • Make sure those reporting to you understand your inter-actions with your mentee (e.g., mentees may drop in at anytime or have a set [scheduled] time to meet with you).
Goal-Setting	• Set clear, achievable goals with your mentor's *assistance*. Remember that the ultimate responsibility is yours. • Continuously evaluate your goals, milestones, and aspirations with your mentor.	• Be realistic in your assessment of your mentee's skills and recommend strategic plans to help her to move forward in her career. • Listen to your mentee's personal and professional career goals and provide input within the scope of your experience and expertise.
Accountability/Boundaries	• Keep your word regarding any follow-up activities agreed on during mentor-mentee meetings. • Have healthy and honest discussions with your mentor about what he or she is actually able to assist you with (e.g., job opportunities, career connections). • Do not put your mentor in a difficult position that requires him or her to compromise his or her job, ethics, etc. • Be aware of the limitations of your mentor in terms of his or her expertise and experience.	• Set some well-defined goals regarding what activities you will work on with your mentee. • Ensure that you are able to do what you promise your mentee (e.g.. providing introductions, creating opportunities). If there are unseen obstacles to comple-tion, keep her informed. • Keep your own professional development on track. Do not jeopardize your career in an effort to mentor others. • Do not hesitate to refer your mentee to someone else, if appropriate. Welcome any new relationship that may form. The ultimate goal is the mentee's growth.

Continued

Table 5.1   Expectations and responsibilities of mentees and mentors—cont'd

	Mentees	Mentors
Transitions	• Continuously evaluate the health of the mentoring relationship. Is it helpful, supportive, and affirming or constraining, burdensome, and destructive? • If/when the time comes to leave this mentoring relationship, do so with grace and professionalism.	• Continuously assess the effectiveness of your mentoring activities. Make changes in the frequency and types of interactions, as needed. • Accept the reality that mentoring relationships will change. As far as possible, maintain a positive relationship with your mentee and support her development. That is your *legacy*.

Communicate Clearly and Effectively

A critical factor in the success of a mentoring relationship is clear, open communication. Many excellent books have been written on effective communication, personal styles, and the like, that can help you to understand the dynamics and develop your skills. It is important to understand differences in styles are not necessarily detrimental to the relationship. They, in fact, prove to be more beneficial. For example, if you tend to be very shy, you may need someone who is more outgoing to engage you in the discussion and help you to examine the heart of the topic or issue. On the other hand, if you tend to be assertive and rather definite yourself, it may be most helpful to have an mentor who will listen, ask incisive questions at key points, and offer input only when asked. In the final analysis, only you can decide who is the best match for you.

Honesty and integrity are best. You need a mentor whom you can trust completely with issues related to your personal and professional life and who has the same confidence in you. You want *honest* feedback on your performance, career aspirations, and the potential for your success. This is not to suggest that you must confide deeply personal aspects of your life or that it is appropriate to ask the same of your mentor. Rather, it is to acknowledge the fact that aspects of our personal and professional lives are sometimes entwined and, in the course of the mentoring process, you may need to reveal aspects of your personal life that you do not want to become common knowledge.

Manage the Transitions

Another aspect of the mentoring relationship that requires careful management is your transition from mentee to colleague. Being able to recognize when the relationship has reached its full potential and is becoming less important in your life and being willing and able to "let go" when the time is right require insight and experience (the chapter "Transitions" discusses the process more thoroughly). Your mentor, too, may have difficulty accepting that their advice and support is not needed as much.

Letting go may occur at a logical juncture in your life (e.g., when you finish your postdoctoral fellowship and leave the community to accept a research position in industry) or result from a gradual drifting apart (e.g., when your research focus changes and you no longer attend the same annual meetings). You may also plan the separation through strategic decisions. For example, you may choose to apply for a research grant in a related (but different) area than your mentor's or seize an opportunity to lead a new initiative at your company, without first consulting your mentor. At the same time, your mentor may be letting go of the relationship by encouraging your interest in becoming involved in new collaborations in related areas or by supporting your application for a position that is completely separate from anything the two of you have been involved with in the past.

Often, the greatest challenge in the transition is with timing: mentee and mentor may recognize the need for a separation but at different points. A mentee may become overly dependent and may feel paralyzed by the very thought of making a major move without the mentor's input. On the other hand, a mentor may be offering too much unsolicited advice or trying to have too much control over the mentee's decisions. If either one begins to feel uncomfortable about the relationship, it is important to talk about it openly. Perhaps, through the discussion, they may see the value in continuing the relationship and can negotiate a new way of interacting. But if they both realize that the relationship has run its course, they may be able to work out an amicable way of moving forward. No one likes to feel that they have been rejected or that their time and efforts have been wasted. If the separation is handled properly, both parties can move on in a healthy manner, actively pursuing science and thriving at the same time.

BECOMING A MENTOR

Traditionally, men who are members of the dominant culture have had *immediate* access to the informal networks and mentoring opportunities available within their workplaces—because of their gender and status—

while women and other underrepresented minorities have not. This is slowly changing, thanks to the initiatives, programs, and actions of individuals, institutions, professional associations, and governments. As this kind of significant, organized effort continues, *all* aspiring scientists will have the benefit of mentoring, throughout their careers. And this can only enhance their likelihood of success, increase scientific and technological discovery, and, ultimately, strengthen society. One way to contribute to the creation of a positive future is to become a mentor yourself.

There are several options available to you. You can become involved in one-on-one mentoring by participating in any of the mentoring programs or initiatives that may exist in your own context or by taking the initiative to establish mentoring relationships with individuals whom you are inspired to help. Interactions with your mentees can include one-on-one professional support, individual career coaching, undergraduate research advising, graduate student advising, and personal decisions counseling.

Direct mentoring requires confidence and maturity. A certain amount of risk is involved because you are making a commitment to a person (and she to you). No one can anticipate how the relationship will develop or what issues may arise during your meetings. You need to be able to handle these with professionalism and discretion. Nor can you know beforehand the level of support that a mentee may require. It may be exactly what you can offer or it may be more than you can handle at the current stage of your career. Keeping in mind the expectations and responsibilities of each member of the relationship and particularly your responsibility to establish and maintain reasonable boundaries is critical.

There may be opportunities within your organization or community to learn more about being an effective mentor. Participating in workshops or programs specifically designed to help you develop your mentoring skills is an excellent way to increase your confidence and establish a network of mentors with whom to share experiences and support.

Another approach to mentoring is to actively participate in and/or develop broader-based mentoring programs that target specific groups of individuals. These may become part of the fabric of an organization or society and are worthy contributions to the mentoring enterprise. For example, Dr. Elizabeth Cannon (one of the authors of the chapter "Personal Style"), in her capacity as 1997–2002 National Sciences and Engineering Research Council/Petro-Canada Chair for Women in Science and Engineering, developed an innovative e-mail mentoring program for girls 11 to 18 years of age (SCIberMENTOR) that she continues to lead.

Again, the key is to be sure that you are comfortable with the level of mentoring that you agree to provide to any particular group or individual. Overextending yourself will not help anyone. Likewise, trying to "save the world" through mentoring presents its own set of challenges and frustrations. To be effective, start slowly and modestly. Investigate the opportunities, identify the areas in which you would like to have an impact, and choose what is most appropriate. If, for example, you were interested in precollege mentoring, you could begin by tutoring in an established program at a local community center or teaching a weekend math class with a prepared curriculum once a month. If your goal were to develop a network of women at your company, perhaps a monthly meeting at a local restaurant (announced through a simple e-mail message) would help you to assess the level of interest and potential scope of the final program. But if, in the former examples, you were to initiate a daily, after-school program in a community that does not have a large enough population of students in the age range that you are targeting or you managed to convince senior management to sponsor a companywide symposium on women's issues in the workplace that was not well-attended because a similar event was offered by the local chapter of a professional association, you would be wasting your valuable time and energy. Each of these larger programs has its own merits—at the right time and in the right contexts. But if you pursued them under these conditions, you could compromise your reputation, suffer burnout, and, more critically, deviate significantly from fulfilling the key responsibilities associated with your work, family, and social life.

Obviously, you will need to decide which approach to mentoring best fits your personality, current circumstances, and time constraints (assess both your professional *and* personal commitments before making a decision). It is also important to assess your level of comfort with self-disclosure. For example, the prospect of speaking with a group of middle schoolers about your life as a scientist may be unattractive to you because you know they are apt to ask rather personal questions (e.g., "How much money do you make?"). On the other hand, speaking about family issues with a small group of early career scientists at a networking session sponsored by your professional association may be just the opportunity that you are seeking to share some of your own experiences of balancing professional and personal life.

The most important thing to remember is to mentor by *example*. You may not have a lot of time to work with an individual or an organization, so in every situation, let your own behavior be the model. During a time when you have great family responsibilities, your formal mentoring involvements may decrease to the occasional lecture at the local girls' science club meetings. Your presence in the field—as a student striving for your first, second, or third degree; as a practicing scientist; as an accomplished mathematician; and as an engineering faculty member—is proof positive that women *can* accomplish much in the fields of science and engineering. Your legacy will be established through your own achievements as a scientist and through the lives of the successful people whom you mentor. As they move forward in their own careers, contributing to science and society through their work and by becoming mentors themselves, they will never forget their own mentors: the women and men who positively affected their lives and helped them to make their dreams become reality.

REFERENCES

JustGarciaHill. [Online]. Available at: http://jgh.hunter.cuny.edu (accessed August 20, 2005)

MentorNet. [Online]. Available at: http://www.mentornet.net (accessed August 20, 2005).

PATRICIA RANKIN, Ph.D.
Associate Dean for the Natural Sciences
University of Colorado, Boulder, UNITED STATES
Physicist

Dr. Patricia Rankin's fascination with understanding the scarcity of antimatter in our universe parallels another professional interest: addressing the underrepresentation of women in administrative positions in science, mathematics, engineering, and technology. A professor of experimental particle physics, Associate Dean for the Natural Sciences at the University of Colorado, Boulder, Department of Energy Outstanding Junior Investigator Award winner (1998), and Alfred P. Sloan Fellow, she was instrumental in attracting $3.5 million from the National Science Foundation (NSF) to strengthen the leadership skills of all faculty (regardless of gender) and increase the number of women scientists in leadership roles.

Born in Liverpool, England, Dr. Rankin attended an all-girls school (from age 11 to 18) where she learned, among other things, "how to make a proper cup of tea" and that women could do anything. When she arrived at university, she was surprised to discover that physics was not a popular subject for women to study. Nevertheless, she enjoyed her time at Imperial College, London, and is eternally grateful to the professor who told her it didn't matter what she studied—what mattered was what she learned. She thinks less fondly of the gentleman who on discovering she wanted to be a particle physicist, remarked that this would be a good career for her because she could always get married if she weren't successful.

After completing her Ph.D., she accepted a postdoctoral position at the Stanford Linear Accelerator Center and made the United States her permanent home. She earned tenure ahead of the usual time frame, and then worked for two years in Washington, DC, as a program officer at the NSF. Returning to Boulder, she successfully applied for an NSF "ADVANCE" award to transform the University of Colorado. She is now the Principle Investigator of the Leadership Education for Advancement and Promotion (LEAP) program and is happily working toward the achievement of her goals as the Associate Dean for Natural Sciences.

●●●

JOYCE McCARL NIELSEN, Ph.D.
Associate Dean for Social Sciences
University of Colorado, Boulder, UNITED STATES
Sociologist

●●●●●●●●●●●●●●●●●●●●●●●●●●●●●●●●●●●●●●

At first glance, the seemingly unrelated research topics explored by Dr. Joyce McCarl Nielsen in her 30-year academic career suggest a dilettante. From river running in the Grand Canyon (to determine the sociological carrying capacity of wilderness areas) to gender stereotypes, battered women, recycling, and social exchange basics, her work has employed almost every viable social science research method: small group laboratory experiments, surveys, ethnography, direct social intervention, focus groups, field studies. But there is a common thread running through this cacophony of projects: a commitment to methodological rigor and the study of issues related to social inequality. These same interests are reflected in her service to the University of Colorado, Boulder: Chair of the Faculty Council Committee on Women, Chair of the Human Subjects Committee, Women's Studies Director, Chair of the Department of Sociology, and currently Associate Dean for Social Sciences.

A young scholar in the late 1960s and early 1970s, Nielsen was strongly influenced by the women's liberation movement. As a graduate student, she cotaught the first "Sociology of Sex and Gender" course offered at the University of Washington, then moved to the Department of Sociology at the University of Colorado, Boulder, where she continues to pursue her two loves: the study of gender and environmental sociology. She helped define the field of gender studies with her texts *Sex and Gender in Society: Perspectives on Stratification* (1st edition, 1978, Wadsworth Publishing; 2nd edition, 1990, Waveland Press) and *Feminist Research Methods: Exemplary Readings in the Social Sciences* (1990, Westview Press). She currently is a coprincipal investigator and active researcher on the University's NSF ADVANCE LEAP grant, headed by Dr. Patricia Rankin.

Chapter

6

NETWORKING

Patricia Rankin and Joyce McCarl Nielsen

Networking is not about schmoozing our way to the top, but about establishing connections with people with whom we share interests. Developing a network is essential in science. It will support you through challenging times and be instrumental in advancing your career, as you'll read in the following story.

> Two years into her graduate program, a struggling Ph.D. student began to realize that her escalating discomfort and unhappiness lay not in the demands of the academic work, the rigors of the laboratory, or with the standards or expectations of the department, but rather in the perpetual tension between her supervisor and herself. Their personalities and styles were simply incompatible. She tried everything to reconcile the situation but finally had to admit to herself that she simply could not continue her Ph.D. with this supervisor.
>
> Embarrassed and humiliated, she felt like a failure and believed that her only option was to quit. But through the support and encouragement of trusted friends and colleagues, she developed the confidence to approach the Chair of Graduate Studies about changing supervisors. With his help, she was accepted by a new supervisor and began a new research project. She felt shaky at first, expecting at any moment to face the same conflicts that she had before, but in time, and with the reassurance of her friends and new supervisor, she regained her confidence. She worked diligently and began to see some exciting results. Three years later, she has four publications in top-flight

journals and three more in preparation or in press, has presented at international conferences, and is now writing up her thesis. Through her own networking efforts, she has secured an excellent postdoctoral fellowship working on cutting-edge science.

WHAT IS NETWORKING?

A network is a group of people who are connected in some way and who exchange something, such as information, ideas, or favors. You can think of everyone you know as being part of a giant network or of yourself as a hub connected to several different networks that govern different aspects of your life. Networking implies communications that flow to you and from you. The balance in communication flow will fluctuate; as you progress in your career, you may respond more frequently to the needs of others or find that you are expanding your network in some areas to meet new needs while maintaining existing links. Generally speaking, the more people you know, the more opportunities you will have to benefit from your network. Your chances will improve even more if your network is made up of people with a broad range of interests. You will not be in contact with everyone in your network at all times and not everyone in your network will be helpful in dealing with any one situation you find yourself in.

Being able to network is not something that you must be born with. Like any career skill, you can learn it. In fact, networking is probably easier to learn than calculus. And if you practice enough, observers will probably think you do it naturally. Anyone whose skills you admire probably is practicing and polishing them whenever they can.

You may be networking without consciously knowing you are doing so. If you are an active member of a community, for example, you are networking. If you cooperate and participate in neighborhood activities, you are networking. If you take someone's class when he or she has to be out of town and he or she returns the favor, you are networking. If you are developing a list of people whom you can call on in an emergency and who will be able to call on you in turn, you are networking.

There are several excellent guides for women in the sciences and engineering that cover a broad range of skills that will make you effective at what you do (e.g., Barker, 2002; Williams and Emerson, 2002), and several sources of good information directed specifically toward faculty members at institutions where research is important (Wankat, 2002). All identify *networking* as one of the most important skills that you will need in your career—whatever your position or responsibilities. If you are just interested in knowing how to network more effectively, this chapter will help, and you can skip directly to the section "Mentoring Versus Networking." If you do not think that you should have to network, that you do not need to network, or that networking is an inappropriate way to behave, read on.

Networking is Trading: The Norm of "Reciprocity"

It may concern you that participating in a network will require something of you, in return for what you receive. This is correct: networking is a form of "social exchange"; it involves *interactions* rather than a one-way flow of favors. While you need not know anything about the theory of social exchange to network, your efforts may be more effective if you can identify general patterns and understand and interpret specific interactions. Social exchange in most societies is governed by what Alvin Gouldner (1960) called the "norm of reciprocity." It means, in essence, that when someone does something for you, you are under some obligation to reciprocate in kind (i.e., with something worth about the same amount or of equal value). This is an unwritten, taken-for-granted norm rather than an explicit rule. If you reciprocate inappropriately (e.g., give someone a new car in response to a dinner party) or bluntly state your intention to repay your debt (e.g., at the end of a nice dinner party, you say something about inviting your hosts back to repay them), you have violated the norm of reciprocity. The extent to which you can be obvious about reciprocating, the time delay between receiving and reciprocating, and the understanding of what is an appropriate good or service with which to reciprocate differ from society to society and with one's place in society.

Establishing a sense of trust—so critical to networking relationships—takes time and rarely is explicit. The longer the interval between specific exchanges, the more trust there is in the relationship (i.e., each of you is confident that your help, support, etc., will be reciprocated, some time in the future; you do not usually remind each other of the service rendered). Note that industrial societies increasingly have less trust built in to social interactions than nonindustrial societies, insofar as there is usually no waiting period between receiving and paying back. For example, when you buy something, you have to pay for it (or promise to) immediately. These more formal interactions leave no room for social exchange, for reciprocation, and hence, there is less room for trust.

You may be concerned that the obligation that you assume in a networking exchange will require you to reciprocate in a way that may make you feel uncomfortable, but the norm of reciprocity is such that it is more likely to be balanced trade. If you feel more is expected of you than you received, then this norm probably is being violated; you can feel comfortable about declining the request. Also keep in mind when networking that the return in kind may not be directly to the person who helped you (e.g., your mentor) but to someone else instead (e.g., someone you mentor). This is a modification of the "norm of reciprocity" that, likewise, benefits society and the individuals involved. Finally, if you cannot or do not want to participate in social exchange in any given society or organization, you will not be able to benefit from being part of the trust network. This will increase your likelihood of missing opportunities for advancement.

Is Networking More Difficult for Women?

There is no evidence that professional networking is more difficult for women. The gender pattern[1] for social contacts actually *favors* women since they tend to be more skilled at making social contacts and are usually willing to put more effort into their relationships (and both assist in building an effective network). Some authors argue that women should be at more of an advantage in today's workplace (e.g., Helgesen, 2001) precisely because they tend to be more skilled communicators and relationship builders.

Network Composition

Research shows that the personal networks of men and women differ in composition (Moore, 1990), with women's including more family members and men's tending to include more coworkers. The same research reveals that these gender differences arise partly from the fact that the roles of the average man and the average woman tend to differ and, likewise, their positions in the social structure. Differences in position lead to differences in opportunities for making certain kinds of contacts and ties and also lead to constraints on personal ties. The reassuring news is that when employment status, family status, and age are the same, gender differences are reduced. Some do remain: women still have more and greater diversity of family ties compared to men in similar positions in the social structure.

---

[1] We use the phrase "gender pattern" in this chapter to highlight any behavior that one gender is more prone to than the other. Gender patterns are not destiny; not everyone of a particular gender will show a particular behavior. Even if you do follow the patterns for your gender, you can learn how to break out of them if needed. This concept is related to that of "gender schemas," which describe the assumptions that we tend to make about how people of a certain gender will behave (see Valian, 1999).

Use of Networks

Women are likely to use their networks in different ways and for different reasons than do men. While not true of everyone, it is well established that women tend to seek support and collaboration rather than challenges and competition (Gilligan, 1982). They are more likely to be concerned about issues related to balancing professional and personal life and about how to juggle multiple, conflicting demands on their time. Many junior women in academia are interested in discussing the timing of having a family. Sometimes women just want to know that their experiences are not unique.

Integration into Informal Elite Networks

Network analyses examining people in formal positions of power show that men are more integrated than women into informal elite networks. Women still remain "outsiders on the inside" because they are not part of the "male culture" (Moore, 1988). Men share common understandings of the rules, styles of competing, and bartering that women are not always exposed to. Heim and Murphy (2001) explore how women need to adopt the rules that work for men to be effective, but even when women do learn these rules, they may find that the rules do not work the same way for them—this makes it all the more important for women to become more proactive about their networking, particularly early in their careers.

IS NETWORKING REALLY NECESSARY?

The opportunity to contribute to this book arose when the editor contacted a colleague of ours who was able to connect her to us. The editor used her network to advance her plans. Part of our preparation for writing this chapter was to send out requests for anecdotes about networking to illustrate the points that we make. No one contacted us to say that they had had a bad experience networking or that they regretted networking. No one told us that they had not benefited from having a network or that it was impossible to develop one. We have not been able to find any article in the literature that suggests that networking is ever detrimental to a career. One research study did identify a rare circumstance in which networking does not offer a significant advantage to career advancement (Requena, 1991). The author found that, in Spain, where there is a "rigid bureaucratization of...employment," networking is relatively unimportant. This suggests that when the process of advancement is highly structured and completely transparent, access to the information relevant to climbing the ladder is available to all equally. But this is no reason to forgo networking activities, if you work in such a regulated environment, for there are other benefits to networking (described in the following sections).

Networking Works. So why doesn't everyone network?

Networking Helps People Know You Better

Sometimes people resist the notion that they need to network because they believe that the activity is only about being "nice" to influential people so they will provide assistance in getting ahead. Another frequent response is that networking should not be necessary because people should be able to advance on their own merits. First, networking is about being pleasant with everyone—including those we perceive to be important—and establishing relationships with people with whom we have something in common. We need to treat *everyone* with respect, regardless of their role in the organization. Second, being promoted on merit requires that your career goals and your merit are *known*. One lesson that you'll learn from networking with more experienced people is that it is usually a mistake not to discuss your accomplishments, hopes, and expectations. To repeat Nadia Rosenthal (from the Prologue): "Male or female, you won't get what you do not ask for in this world." The deeper issue here, though, is the fact that women in the sciences and engineering often are resistant to the idea that it is appropriate and necessary to manage their careers.

Networks Help You Stay Connected

Networks help you stay socially involved and integrated into the day-to-day activities of your professional and personal life. Sometimes this may seem like a waste of time: why spend even a few minutes talking to someone when a deadline is fast approaching? It often is easier to justify making time for things that are clearly and directly related to our career goals than for what appears to be an optional activity. Especially in the early years of establishing your reputation (and particularly in North America), you probably are working long hours and are feeling pressured to produce results. However, taking a few minutes to exchange pleasantries with someone in the laboratory or hallway can provide an important opportunity to step back a bit and reflect on what you are doing. Who knows? The person with whom you are speaking may suggest something important that you had not thought of. Often just talking about what you are doing can help to clarify your thoughts; other times, it can lead to a new breakthrough. The point is that the process of interacting—in and of itself—is valuable. And short breaks such as these will help you avoid burn out (as you'll read in the chapter "Mental Toughness").

An interesting study was published recently about people who thought they were lucky and people who thought they were not (Wiseman, 2003). The author suggested that "luck" is more a matter of being open to opportunities than a matter of random fortune. Networks help provide these opportunities. One way is by encouraging the sharing of information and resources that can lead to important scientific breakthroughs. You can enjoy significant professional benefits from being able to tap into information sources that can provide access to specialized equipment and advanced information on new techniques that are being developed. Similarly, you can benefit from brainstorming sessions and exchanging ideas with a broader group of people.

Networks Provide Support

Sometimes what you seek from members of a network is not suggestions about what you can or should do, but support and/or affirmation. This is an equally valid use of your network. We have all been in situations that trigger strong reactions that we do not express because we wish to maintain our professionalism by responding appropriately. For example, how do you deal with the feelings engendered by an insensitive comment of an influential (and probably well-meaning) person who has just told you not to worry about earning tenure because "You are one of the rare women in your department; they would never turn you down"? or with the annoyance resulting from the complaints of absentee students who claim that, by "failing them," you are "ruining their future"? Politely. Then let go of the emotional energy by venting to a trusted member of your network. (The chapter "Mental Toughness" suggests other strategies for defusing tension and letting go.)

It's Acceptable to Ask for Help

You may agree that it is acceptable to manage your own career and that networking would help but still not take action. Scientists and engineers are trained to be independent problem solvers and place a high value on individuality and taking personal responsibility. As a result, you may feel that if you cannot handle every situation presented by your job—on your own—you are in the wrong position (an example of emotional reasoning, described in the chapter "Mental Toughness"). And you may believe that asking for or accepting help—rather than struggling through everything all by yourself—is a sign of weakness. If you do, you are mistaken. The fact is that this feeling of being in a position that you are not entirely qualified for is surprisingly common, even among successful people. There is even a term—"impostor syndrome" (Clarkson, 1998)—that has been coined to describe this attitude.

Realize that asking for help will not imply that you are incompetent or inappropriately placed in your job. It is, in fact, a sign of competence to be able to recognize when you need the help of others; you will not receive credit for wasting time and resources on a problem that could easily be solved through consultation. With the long-term trend in science toward increased collaboration and teamwork, you are expected to take advantage of the expertise of others in the group. Focus on the fact that there is a large difference between asking for advice and asking someone to do all the work. An effective network is a lever that will increase what you can achieve rather than take away from your capacity.

Networking is an Ethical Way to Advance Your Career

Another concern shared by many is that it is somehow manipulative or unethical to use your career management skills to your advantage by developing a better understanding of your work environment and the hidden agendas, rules, and regulations. This belief is so common that it is worth probing. We suspect that attitudes toward career advancement strategies obey a gender pattern; that is, that women are more likely than men to worry about how appropriate it is to make the advancement of their career part of their job. Is it wrong to be an effective persuader? Is it wrong to use your skills to make the environment better for everyone? One book that probes the issues about how ethical it is to be a leader is *Political Savvy: Systematic Approaches to Leadership behind the Scenes* (DeLuca, 1999). DeLuca makes an important distinction between doing what is best only for yourself and doing what is best for the whole. The former is Machiavellian; the latter is leadership. Of course a balanced approach is necessary; you want to avoid the mistake of always putting the organization first and never taking your interests into account. If the best idea was always accepted and the best person awarded the job, we would be living in a very different world. Until then, our advice is to learn how to improve the chances in favor of getting what you want for yourself *and* society.

We all need to think about whether we are exploiting our assets as effectively as we can—Are you reluctant to interpret relationships as assets? Research indicates that being able to access advice helps women advance their careers (Sonnert and Holton, 1995) and that networks often have an impact beyond the education and skills possessed by an individual (Lin, 1999). In the experience of one scientist:

> *I do think networking is invaluable. When I have a strong network, my career progresses much more smoothly. I believe that I have more information on which to base my decisions, and I feel more confident in my choices and in the "fit" of the positions I accept. Many jobs (especially postdoctoral positions) and other opportunities (like serving on a grant review panel) are filled before they are advertised, and it's through your networks that you learn about them. I do think that women often have more limited professional networks and therefore have more difficulty tapping into these "hidden" jobs. Also I believe many are not aggressive enough about forming and building networks, mostly because we do not appreciate how invaluable they can be. I found it to be particularly important to have a wide network as I shifted out of an academic physics position into medical physics.*
>
> Kristi R. G. Hendrickson, Senior Fellow, Radiation Oncology,
> University of Washington

The breadth and effectiveness of your network will directly influence the ease with which you can attain your goals. Through your contacts you can obtain information more specific to your needs, and faster, than going through traditional "channels." Instead of searching for specific policy documents on the Internet, for example, you may find that you can obtain the most relevant, up-to-date information—and useful advice on how to take the best advantage of it[2]—from a colleague. More importantly, if you have an effective network, you may be privy to opinions that may not otherwise be available or forthcoming, such as whether a certain individual would make a good supervisor or whether a particular location would be an enjoyable place to spend a sabbatical.

Networking Teaches You the Rules of the Game

Much of the recent work on the advancement of women argues that the way to change institutions is to increase the number of "subversive insiders" or "tempered radicals" (Meyerson, 2003), people who can work from within the system but who do not accept it as perfect. It is reasonable to be concerned about becoming part of a system that you do not endorse, but

[2] For example, queries about "tenure clock" policies may lead to discussions about the implications of taking time out to have a family, what you should be aware of (in the context of this particular employer), and what other factors to consider before making a decision.

choosing not to participate is rarely the most effective strategy for making it better. We think that to improve the working environment for everyone, we need a greater "unobtrusive mobilization" (Katzenstein, 1990) of people who have a broader perspective on and understanding of the issues relating to equity and professional and personal life balance. We would like to see more of these individuals move into positions where they can act on their ideals.

However, as Gail Evans discusses in her book *Play Like a Man, Win Like a Woman. What Men Know about Success That Women Need to Learn* (Evans, 2000), before you can change an organization, you need to succeed in it long enough to obtain a position of power. To be successful, you need to understand how the organization works. It is essential that you know the rules that govern the organization and how to use the rules to your advantage. You have to know, for example, what determines who gets promoted in an organization; is it team playing or personal initiative? Do not misunderstand—we are not saying that these are the "right" rules or that they should always remain, just that to change them, you need to know what they are (at the very least). We are, after all, working within an imperfect system. Life is not always fair. Having acknowledged that, only you can decide how to respond to the environment in which you are currently. Since you are reading this, we assume that you prefer to be·proactive.[3] Again, we stress the importance of attaining a position of influence if you want to shape an organization.

It may sound as though we're suggesting that your career is a game. Certainly the "life is a game" metaphor is one that is used frequently. However, we are talking of a competitive game, not a collaborative one. This is a game in which the results matter. It's a game that you can and need to train for. You study to master the technical skills for the work you do and to stay current. Accept that you need to learn a broader range of career skills—such as networking—to ensure your success.

MENTORING VERSUS NETWORKING

We are not advocating replacing mentors with networks. Mentors are important, especially in the early stages of your career, when your thesis advisor often plays this role. Traditionally, mentors are people who have much more experience and influence than you do. You will often think of them as people who are likely to have your best interests at

---

[3] Proactive doesn't mean always active against real or perceived wrongs. We encourage you also to be strategic and use your network to help you to decide what battles to fight and when.

heart. In fact, your mentor may well be the first member of your network to support your career and provide advice on what you should do. However, research indicates that in male-dominated fields, the predominant style of mentoring tends to be a better match with a "male" interaction style. This style tends to focus on providing specific information or advice on technical issues and usually the primary goal is for the protégé (mentee) to establish his or her independence from the mentor. Indeed, some authors have discussed "the heroic mentorship" (Broome, 1996) as one that ultimately leaves the protégé alone to slay the dragon.

A very good discussion of the different styles of mentorship can be found in Chesler and Chesler (2002). Collective mentoring, for example, makes the guidance of any junior members of an organization the responsibility of all the senior members of that organization. This team approach is clearly a good idea in principle, but one that is rarely found operating in practice, especially in the professions. Multiple mentoring, another model, encourages people to find several people with disparate skills to act as their mentors. Packard (1999) talks about assembling a diverse group of mentors into a "composite mentor." This approach has the advantage of encouraging people to think about their specific needs for mentoring and to look for people who can provide advice on those specific needs. Building a composite mentor will connect you to different groups of people, just as effective networking will. (The chapter "Mentoring" explores the topic in greater detail.)

So, what is the difference between having a network and having a composite mentor? Mentorship implies a teacher–student relationship. Mentors have the knowledge. Mentees want and need to learn it. Networking is more of a reciprocal relationship—information flows both ways. Sometimes you will benefit from it, sometimes you will be helping someone else. While networking looks very similar to the multiple mentorship model, it is much less hierarchical in structure. Networking also encompasses more casual contacts. As an effective networker, you will be part of a wider support structure and receiving advice from a broader range of people. You will have to decide what advice to take and learn how to pick and choose between conflicting suggestions.

Finally, there is peer mentoring or co-mentoring. Peer mentoring refers to mentorship by people who are very close in age and/or experience to those whom they are mentoring. First-year graduate students might be matched with second-year students, for example, to help them navigate the transition from undergraduate life. There are few, if any, real hierarchical relationships between participants in a peer mentor relationship. Networks, however, can involve people in a variety of positions within an organization or outside of it. Again, mentoring and networking are not mutually exclusive activities; each provides you with a range of opportunities from which you can benefit.

NETWORKING STRATEGY

Build on the Networks You Already Have

You have probably spent much of your life networking without realizing it. When you were growing up, your network consisted of your family and friends; at college or university, you may have belonged to study groups[4] that formed part of your network of classmates, faculty, coop, or work colleagues and friends. Each and every interaction in your daily life is a potential networking opportunity, from working with a student to introducing yourself at a meeting. Be aware of the possibilities and prepared to take the initiative when you meet someone who also is interested in networking.

The following stories illustrate how seemingly casual contacts can result in expanded professional opportunities and increased funding support.

I once received research funding from a nonprofit organization that approached me with a research project they wished to have implemented. It was an opportunity that presented itself to me as a result of a networking relationship I had developed years before. A former student with whom I had become friends started a business next door to this organization and had made the acquaintance of the Director. When my friend learned of their vision for developing online databases to support and change their business model, he immediately thought of my work and recommended they contact me. As a result of this referral, we had two years of funding for three students and learned a great deal about GIS systems and how to integrate them with databases. Our work demonstrated a certain level of expertise in our student group that made us eligible for several practical research projects with geosciences and the university transportation center. This has generated much more funding over time, in the form of student support on multidisciplinary projects.

Some years ago, a former manager in a federally funded laboratory in the Washington, DC, area moved to my state to join his wife who had accepted a new position in the community. He was interested in networking with people in

---

[4] If you are reading this while still at college or university, wonderful! If you are not a member of a study group, you are missing an excellent opportunity. Even if you find the material that you currently are studying to be easy to understand, this is unlikely always to be the case. By establishing sound study habits now, you will reap the benefits later. Besides, explaining things to others will help you to deepen your own understanding.

my department but most seemed uninterested. I accepted his proposal to work together on a research project, but nothing really followed for we did not get the funding. But through our collaboration, he introduced me to deans and researchers in other departments at the university as well as to representatives from industry and the state economic development groups. I was amazed by how easily he could knock on a door, schedule a meeting, and become acquainted with these people. It really opened my eyes! We prepared a presentation with one group for a small conference on technology transfer. We wrote a proposal to start a small "collaboratory" on campus that would facilitate IT work among many partners from different colleges and departments. Since then, people on campus view me as someone who will collaborate. I have been included on many proposals and asked to lead others and am now flush with funding. If I had to identify how this all originated, I would say that it was meeting this accomplished networker who was trying very hard to establish new contacts in his adopted state. I am thankful that he swept me up onto his network and showed me how it is done.

JOAN PECKHAM, Computer Scientist, University of Rhode Island

Your connections don't have to be current for you to be able to establish a network. They may have lapsed for some period of time.

About seven years after graduating from college, I was completing my Ph.D. at the University of Wisconsin, Madison. My first-year roommate, with whom I had lost touch years before, was a relatively new graduate student in the fine arts program at the same institution. As she was walking down a hallway one day, she was surprised to read my name on a set of mailboxes—she wasn't aware that I was in town, let alone on campus. She e-mailed me immediately and suggested we meet for coffee. We found that, among other things, we had more in common than we had a decade earlier. As we talked, she mentioned that she was about to leave her paying-the-bills editing job in engineering and was hoping to find a replacement. I was preparing for my defense and looking for work, so I decided to meet with her supervisor, even though I had little interest in technical editing. As it turned out, her supervisor had a passion for engineering ethics (the topic of my dissertation) and offered to hire me to teach a course on the subject the next semester. He couldn't employ me full-time and knew I needed more work to help support my family, so (unbeknownst to me) he contacted a colleague and urged her to accept my application for a research position in engineering that he knew I had recently submitted. That was nearly eight years ago, and through a combination of conscientious work and continued networking, I am now Assistant Dean in that very college of engineering. Thank goodness for both serendipity and networking.

SARAH PFATTEICHER, Assistant Dean of Engineering,
University of Wisconsin, Madison

As these stories illustrate, networks can develop from casual and accidental contacts as well as from deliberate ones. The real difference between successful and unsuccessful networking is that in the former, you are an *active* rather than a passive player in the connection game; you are not relying purely on chance but are continually looking for ways to extend your network and taking full advantage of any opportunities to do so.

If you wish to advance your current career, pay special attention to making contacts with people in your field or related fields. For example, you may choose to join the professional organization(s) for your discipline, attend meetings, or join associations that are related to your work but not specifically in your area. Take part in annual conferences (see section below). Serve on review panels. Invite people in your workplace to join you for coffee, or have lunch with your coworkers. Attend departmental colloquia and seminars and accept any invitation to share a meal with the speaker after the event. If graduate students normally aren't included, ask your supervisor to invite you; suggest that the department invite and pay for students to attend these dinners. If there is a request for people to spend time with a visitor to campus, volunteer. Again, suggest that your department include students in such opportunities.

Casual contacts often provide unanticipated opportunities to take your career in new directions. If you want a *broader* network, then pay attention to what is going on around you. Are there articles in the paper about someone whom you would like to know? Do you have the time and interest to become involved in volunteer organizations? What about joining your alumni association? Do you speak with the person next to you on a plane? Do you have conversations with people in waiting rooms or in queues? Any of these activities could result in unexpected opportunities.

Actively Develop New Contacts

So how do you take a more active approach to networking? The same way that you would form a study group or develop a new mentoring relationship: identify people who may be a good "match" for you (e.g., have similar interests and goals), create opportunities to introduce yourself, judge whether there is potential for a relationship, and, if so, follow through by doing your part to manage and maintain the relationship. (The chapter "Mentoring" discusses strategies as they relate to that activity.)

Identify a Potential "Match"

When you are working on your career network, it is natural to concentrate on identifying more senior people to approach, but do not ignore junior colleagues and students. First, this is appropriate collegial behavior—wouldn't you have appreciated it if someone had introduced you to his or her network? If you are benefiting from networking, make it a point to look out for other people who could benefit from being included and invite them in. Second, you may find that earning a reputation as someone who will help others may attract people who want to work with you. Finally, you do not have to be friends with everyone in your network, but you do need to be able to work with them. You are looking for people with overlapping interests and concerns, not necessarily someone with whom to spend a lot of your time.

Prepare a "30-Second" Introduction of Yourself

Most things are easier to do if you are prepared and practiced. Networking is no exception. Beginning conversations with people whom you don't know will be simpler and more comfortable if you can describe (in a few short sentences) who you are and why you're interested in speaking with them (think of it as a "30-second commercial," if you will). By developing and rehearsing a brief introduction, you'll be past the awkward stage in the conversation before you know it. What is the most important thing that you would like people to know about you and to remember? What do you want to know about them? Tailor your introduction to your listeners. What could you say about your work that would inspire them to ask a follow-up question? Do you have a hobby that someone else might share that could form a basis for discussion? You want people to believe you to be pleasant, but avoid using humor unless you are sure that it will not be mistaken (what people consider amusing often depends on their culture and life experiences). The following examples may give you an idea:

Hello, Dr. Heffram, my name is Jeanne Wayman. I'm completing my Ph.D. in aeronautics with Dr. Cochrane at the University of Toronto, Canada, and would like to do postdoctoral work on hypersonics, focusing specifically on <details>. Your research interests me very much and I wondered if there would be any opportunities in your laboratory?

Dr. Kropinski and I are collaborating on a research project to sequence bacteriophage genomes. We've already completed the analysis of $\phi W14$ and are looking for someone with your expertise to conduct the proteomic analysis. Would you be interested in joining our team?

Create Opportunities to Meet

The success of networking depends, in part, on your being in the "right" places to make the kinds of contacts that you need. Do you know how to approach people directly without introductions from others? Do you attend conferences? Can you take a summer research internship at the place you would ultimately like to work? Has anyone else from your institution gone there whom you could contact for help? With a little thought, you will be able to identify a number of ways to meet people.

Contact People Directly

You do not need to wait for an introduction to someone before approaching them—you can introduce yourself. This is more difficult for some than for others, but there are ways to make it easier. For example, you can communicate with people through e-mail, regular mail, or telephone. Until recently, e-mail was a fairly effective way of contacting people, but with the proliferation of SPAM, this method may be much less successful. If you choose this option, create a subject line that is clear and direct to minimize the chance that the e-mail will be deleted before it is opened. Summarize your message in the heading. For example, "Requesting info on studies of tectonic plates in Pacific" is more likely to elicit a positive response than "Hi." Avoid sending attachments, unless expressly asked to after they reply. You could append your Curriculum Vitae to the bottom of an e-mail message, for example, or (perhaps better) add the sentence "Curriculum Vitae available upon request." (See the chapter "Communicating Science" for more points on "etiquette.") If your request is not urgent, it may be more effective to send a written note, making sure to provide your e-mail address in your contact information.

Such contacts are more likely to be successful if they make a modest request, rather than an elaborate one, and if you do not waste the recipient's time with unnecessary details. Don't bury your request in a long, rambling message. (Have you ever had to listen to a long message on your answering machine just to learn the number to call in return?) Make it clear who you are, what you want, and why you think the person whom you are contacting may be able to help. (This is a perfect time to use your "30-second introduction.") It goes without saying that impeccable spelling and grammar are essential in written communication. This is, after all, the person's first introduction to you, and first impressions can be lasting ones (as you'll learn in the chapter "Personal Style"). Some readers will judge you by your ability to write. Do make it clear that you realize that your request may have come at an inconvenient time and that you will understand if they cannot respond immediately. Know that there is a fine line between following up a request (in case someone did not receive the first one or may have forgotten to respond) and imposing yourself on someone. An effective strategy is to indicate a time frame in which you hope to receive a reply, for example, *"It would be most helpful to receive a reply by the end of next week"* or *"I will follow up at the end of the month."*

Build Networks at Your Institution

Encourage the organizations to which you belong to support networking between members (and by extension, your efforts to network with your colleagues). This can be as simple as providing the contact numbers of other women working in the same business to being more active and hosting gatherings.

> *We have created a group of women faculty in science and engineering that we call "The Network" here at the University of Michigan (UM). We developed a mailing list of every woman tenure track faculty member in science and engineering and invited them to a series of dinners, social events, and lectures on topics pertinent to career development, such as mentoring, work, and family conflict. This has been a huge success. The women have benefited from meeting each other and have forged relationships that cross departmental boundaries. UM is a large, decentralized university, so building a framework for women to connect with one another has proved a key ingredient of our ADVANCE project.*
> Robin Stephenson, Program Manager, NSF ADVANCE Project,
> University of Michigan

One rarely discussed but real advantage of institutional support for networking is this: the more acceptable networking becomes within an organization and the more the administration encourages and supports networking efforts, the easier it becomes to approach people for networking purposes.

If formal opportunities for networking do not exist in your workplace, create your own informal networks.[5] You could identify a topic of general interest and suggest a meeting to exchange ideas, for example, or suggest a discussion while sharing a meal. A group of graduate students could meet regularly to exchange suggestions for applying for a job. Junior research scientists in industry could gather biannually to discuss how to advance within their company. Lobby your administration to help formalize the arrangements and provide meeting space or other resources.

Network at Conferences

Conference networking can begin on the way to the meeting. Introduce yourself to people that you see waiting in the airport lounge reading the materials sent by the conference organizers. Maybe you can sit next to

---

[5] Janis McKenna offered many excellent suggestions on how to network, several of which we used in this section.

them on the plane or share transportation to the meeting. Consider staying at the conference hotel or one of the main hotels rather than looking for the cheapest accommodation you can find—that way you are more likely to mingle with other conference attendees. Wear your name badge and introduce yourself—consider bringing and distributing business cards. Make sure that you attend conference social events—riding in a bus or sitting at a dinner table will provide you with lots of opportunities to meet people. If you are a local or have special knowledge of the area, offer to help out as a local guide. Do not force a contact. If someone does not respond to your overtures, do not continue them—move on and find someone more congenial. Try to avoid only mixing with people whom you know. If you are in a group and see someone who seems to be isolated, welcome them into the group. Go to the talks that interest you (and the ones that are drawing a large crowd), ask questions, and approach the speaker afterward with follow-up questions. If you are junior, consider becoming a session secretary; helping to organize a session will give you a chance to meet with the speakers. One of the authors once worked as a scientific secretary and as a result rode in a taxi back to the airport with a Nobel Prize winner who regaled her with what he had learned during his career. She didn't even have to ask for his advice.

Access Remote Networks or Newsgroups

Thanks to the increasing acceptance of the importance of networking and the increased ease of keeping in contact electronically, it is now possible to subscribe to several online networks tailored for women in science and engineering. These include (among others) the Association for Women in Science[6] and the International Network of Women Engineers and Scientists (INWES).[7] INWES was created following the UNESCO World Science Conference held in Budapest in 1999 in response to a resolution encouraging the development of a "global network" of women engineers and scientists. Organizations such as these often host events in various regions—in addition to regular conferences—moderate online discussion, publish newsletters, etc., that foster networking between and among women.

Several sites focus on women in academia. Though many are aimed primarily at women in specific geographic regions, they often have useful

---

[6] http://www.awis.org. Web site for the Association for Women in Science.
[7] http://www.inwes.org. Web site for the International Network of Women Engineers and Scientists.

information that can be applied generally. The National Initiative for Women in Higher Education,[8] On Campus with Women,[9] Canadian Association of University Teachers,[10] Canadian Federation of University Women,[11] and the American Association of University Professors[12] all have potentially valuable information and are worth a visit. It is also worth checking to see whether any of the professional societies of which you are a member support networks or newsgroups for women. Many do, but if they don't, you can ask them to set one up. Also check to see whether your organization is a member of "MentorNet,"[13] which is set up to provide opportunities for external e-mentoring and includes the option for junior faculty to be matched with tenured faculty mentors.

Remember that different types of contacts and networking efforts will serve different needs. While remote networks are very useful for providing access to people with specific areas of expertise, for example, they are no substitute for local networks. Having direct social contact with the members of your network is an important advantage that cannot be underestimated.

Manage and Maintain the Relationship

Once you have identified people who are potential contacts, introduced yourself, and determined that they are interested in networking, you need to "follow through" appropriately. As in any relationship, clear, effective communication is essential. Clarify your expectations, needs, and goals. Treat people with dignity and respect their boundaries. Reciprocate (see the "norm of reciprocity" discussed earlier). Be considerate and professional in your approach. (The chapter "Mentoring" has some excellent ideas on this topic that are equally applicable to networking activities.)

---

[8] http://www.campuswomenlead.org. Web site for the National Initiative for Women in Higher Education.

[9] http://www.aacu-edu.org/ocww. Web site for the Association of American Colleges and Universities (on campus with women).

[10] Canadian Association of University Teachers (2005) [Online], [Accessed 14 February 2005] Available from the Internet: www.caut.ca.

[11] http://www.cfuw.org. Web site for the Canadian Federation of University Women.

[12] http://www.aaup.org/Issues/WomeninHE/index.htm. Web site for the American Association of University Professors.

[13] http://www.mentornet.net. Web site for MentorNet. The E-Mentoring Network for Women in Engineering and Science.

Minimizing Misunderstandings

One technique for clarifying the terms of a networking relationship (and providing structure for your meetings and conversations) is to create a written description that proposes how and about what you will interact that you send or give to potential members of your network (similar to the information presented in the list of expectations and responsibilities of mentors and mentees, described in the chapter "Mentoring"). This approach is particularly useful (even advisable) in situations in which you feel that there could be misunderstandings.[14] In spite of your best efforts, you may occasionally encounter someone who will misinterpret your desire to network as an interest in a different sort of relationship, not professional, but personal. This situation is less stressful to deal with if the person is a peer rather than someone in a position to influence your career. Regardless, the initial response in either case is to make it very clear, politely but firmly, that he or she is overstepping your boundaries and making you uncomfortable. Emphasize that this is a working relationship, rather than a social one, and that you expect them to behave appropriately. If you can make a statement such as *"I'm sorry, perhaps I did not make myself clear; all I am interested in is...,"* you may help the person to withdraw gracefully and, perhaps, salvage the professional relationship. If you are concerned about potential negative repercussions, make sure to confide in a senior colleague whom you respect and ask his or her advice on how best to proceed.

A related problem is when someone in your network expresses concern about how their relationship with *you* could be misinterpreted by people outside the relationship. There are several approaches to minimizing the chances of this happening, including conducting one-on-one meetings in public places (rather than behind closed doors), making a habit of inviting at least one other person to participate in your networking activities, and maintaining a large network (so you don't spend too much time with this one individual).

Complications to Overcome

Developing a network involves building relationships, and this requires time and energy. If you are balancing work and family responsibilities (as many women are), your response may be that your life already is too busy to add networking to your list of required activities. But making an investment in

---

[14] Trust your instincts here. If you do not feel comfortable working with someone, ask yourself whether you really need to.

networking is similar to making a commitment to exercise: the benefits that it provides (both professionally and personally) more than compensate for the time it takes. Developing and maintaining your network are an essential part of successful career management and can be integrated into your weekly schedule, as you would your other regular activities.

If you tend to be introverted, you probably will find that networking drains your energy, while your extroverted colleagues seem to thrive. Though this difference in temperament will not change completely, with experience networking may become more comfortable and require less energy. Remember that there are many different situations in which you can interact with people; choose those that are naturally more comfortable for you as an introvert, for example, one-on-one conversations in the hall-way, informal, small group interactions at a conference, or larger group activities where you already know many of the people. Promise yourself a reward after the meeting to encourage your full participation or build into your schedule alternating periods of interaction and private time.

HOW EFFECTIVE ARE YOUR NETWORKS?

Social network analysis is a promising new area of organizational sociology (Haveman, 2000). Essentially, it measures the number and strength of net-work ties as well as the distance between them. These data can then be transformed into measures of power, control, influence, prestige, isolation, segregation, stability, and solidarity, among others. An interesting book covering the science of networking is *Six Degrees: The Science of a Connected Age* (Watts, 2002). In it the author describes the "Kevin Bacon game," the goal of which is to connect any given actor to the American celebrity Kevin Bacon. An actor is "linked" to another if he or she has acted with them; that actor is likewise connected to others with whom he or she has worked, and so on until the chain links to someone who has been in a movie with Kevin Bacon. Connecting the first actor to Kevin Bacon can usu-ally be done in six links or fewer.[15]

You can determine the effectiveness of your own networks—albeit in a less sophisticated way than using social network analysis. Create a list of key activities in your life, and under each heading, identify whom you would contact for help or advice. If you have several names in each category, you are doing fine. For a slightly better measure, you can assign a rating to each person based on how well you know him or her and how helpful he or she is likely to be.

---

[15] Robyn Marschke was especially helpful in writing this section.

Strengthen Your Networks

Though you may be most comfortable dealing with people like yourself, your network will be most useful to your professional and personal development if it is broad based and inclusive. It is especially important in the sciences and engineering to understand that male colleagues can be important allies and just as committed to the goal of helping women to succeed as female colleagues may be. Male faculty, for example, who still outnumber female faculty in the sciences, may be able to take more time to help you precisely because there are more of them, and fewer women are taking advantage of them as a resource. Remember, too, that women faculty in fields different from your own are still likely to have experiences in common with you. You are looking for people with whom you can establish a rapport. You also want some people in your network to have skills that are complementary to your own.

> When I was at Boston University (BU), I founded a program for high school girls called the Pathways Program (www.bu.edu/lernet/pathways) that brings together 400 high school girls and teachers over a two-day period and involves over 100 volunteers each year, from inside and outside BU. Many women from all areas of science and engineering at BU—undergraduates to full professors—work together to make this program a reality. Over time, the women who worked on Pathways came to know one another and formed an informal mentoring network. More senior people offered advice about grants and preparing tenure documents to more junior people—even giving the more junior people copies of old grants or tenure documents to review to "show them what a successful one looks like." Students would give tours or come do demonstrations with their professors and have an enjoyable time together talking about science. This was an unusual network in that it crossed disciplines (because there are so few women in any given science department). I was both mentor and mentee in this network and benefited greatly.
>
> ELIZABETH SIMMONS, Professor of Physics, Michigan State University

Ask for Feedback and Assistance

If, in spite of your best efforts, you still feel your networking efforts are failing, seek the advice and assistance of someone whom you know will be absolutely honest with you. Ask for a frank assessment of your networking technique. Make it clear that you value his or her honesty and will not be offended by anything that he or she says. Ask them for feedback on the people whom you are trying to contact (e.g., Are they at too high a level? Already overextended? People who are not the best "match" for you?). Ask whether you are taking the correct approach, if your attitude could be the problem, if there is something that you need to change about your interaction style.

When you find someone who will tell you what you can improve on and/or offer concrete suggestions as to what to do differently, treasure that person! One of the authors once prepared for a job talk by giving her presentation to several colleagues. All but one of them offered flattering and encouraging comments. Fortunately, the one dissenter was willing to point out several major defects with her talk, including the fact that the slides could not be read from the second row.

NETWORK FOR YOURSELF AND SOCIETY

Networking is an activity that is something you can do for yourself that will benefit your career and personal life. You are free to select your level of participation. You may wish to begin modestly: think of one problem that you could use advice on and set a goal of adding one person to your contact list who could help. If you do not feel comfortable asking anyone for advice, then set a goal to get to know one person well enough to feel comfortable doing so.

We encourage you to network because we want you to succeed in science. We want everyone to be as effective as they can be in whatever they choose to do, for we believe that this will help to catalyze changes in the systems in which we work that will make them more inclusive and humane for all. No matter how competent and hardworking you are as an individual, you will achieve more through networking.

> *Never doubt that a small group of thoughtful, committed citizens can change the world. Indeed, it is the only thing that ever has.*
> Margaret Mead, Anthropologist

ACKNOWLEDGMENTS

The authors thank all the people who assisted in the preparation of this chapter by contributing their stories and sharing their thoughts and insights. In particular, we acknowledge Naomi Chesler, Robyn Marschke, Janis McKenna, and Amanda Peet.

REFERENCES

Barker, K. *At the Helm, a Laboratory Navigator*. Cold Spring Harbor, NY: Cold Spring Harbor Laboratory Press, 2002.

Broome, T. The Heroic mentorship, *Sci Commun* 1996;17(4):398–429.

Chesler, N. C., and Chesler, M. A. Gender-informed mentoring. Strategies for women engineering scholars: on establishing a caring community. *J Eng Educ* 2002;91(1):49–55.

Clarkson, P. *Achilles Syndrome: Overcoming the Secret Fear of Failure.* Boston: Element Books, 1998.

DeLuca, J. R. *Political Savvy. Systematic Approaches to Leadership behind the Scenes.* Berwyn, PA: EBG Publications, 1999.

Evans, G. *Play Like A Man, Win Like A Women. What Men Know About Success that Women Need to Learn.* New York: Broadway Books, 2000.

Gilligan, C. *In a Different Voice: Psychological Theory and Women's Development*, Cambridge, MA: Harvard University Press, 1982.

Gouldner, A. W. The norm of reciprocity: a preliminary statement, *Am Sociol Rev* 1960;25:161–178.

Haveman, H. A. The future of organizational sociology: forging ties between paradigms, *Contemp Sociol* 2000;29:476–486.

Heim, P., and Murphy, S. A. *In the Company of Women.* New York: Jeremy P. Tarcher/Putnam, 2000.

Helgesen, S. *Thriving in 24/7. Six Strategies for Taming the New World of Work.* New York: The Free Press (Simon and Schuster), 2001.

Katzenstein, M. F. Feminism within American institutions: unobtrusive mobilization in the 1980's. *SIGNS* 1990;16(1):27–54.

Lin, N. Social networks and status attainment. *Annu Rev Sociol* 1999;25:467–487.

Meyerson, D. E. *Tempered Radicals. How Everyday Leaders Inspire Change at Work.* Watertown, MA: Harvard Business School Press, 2003.

Moore, G. Women in elite positions: insiders or outsiders? *Sociol Forum* 1988;3(4):566–585.

Moore, G. Structural determinants of men's and women's personal networks, *Am Sociol Rev* 1990;55:726–735.

Packard, B. A "composite mentor" intervention for women in science. Presented at the American Educational Research Association Annual Meeting, Montreal, QC, 1999.

Requena, F. Social resources and occupational status attainment in Spain: a cross-national comparison with the United States and The Netherlands, *Int J Comp Sociol* 1991;32:233–242.

Sonnert, G., and Holton, G. *Who Succeeds in Science? The Gender Dimension.* Piscataway, NJ: Rutgers University Press, 1995.

Valian, V. *Why So Slow? The Advancement of Women.* Cambridge, MA: MIT Press, 1999.

Wankat, P. C. *The Effective, Efficient Professor. Teaching Scholarship and Service.* Boston: Allyn and Bacon, 2002.

Watts, D. J. *Six Degrees: The Science of a Connected Age,* New York: W. W. Norton & Company, Inc., 2002.

Williams, M., and Emerson, C. J. *Becoming Leaders: A Handbook for Women in Science, Engineering and Technology.* St. John's, NF: NSERC/Petro-Canada Chair for Women in Science and Engineering, 2002.

Wiseman, R. *The Luck Factor: Changing Your Luck, Changing Your Life: The Four Essential Principles.* New York: Hyperion, 2003.

Chapter

7

. . . .

MENTAL TOUGHNESS

. . . .

Peggy A. Pritchard, Editor

> *Prepare yourself for the world, as the athletes used to do for their exercise; oil your mind and your manners, to give them the necessary suppleness and flexibility; strength alone will not do.*
>
> Earl of Chesterfield

MENTAL TOUGHNESS: THE "PHYSICAL FITNESS" ANALOGY

One of our greatest assets is our mind, yet how many of us consciously train it to serve us? Just as competitive athletes train their bodies for peak performance, scientists can develop "mental toughness" to enhance their professional and personal success and satisfaction. The nature of science itself demands it, the imperfect systems in which we work require it, and our own natures thrive on it.

Mental "toughness" is not about being unfeeling, uncaring, or impervious to criticism and disappointment; it is not about being someone we are not. Rather, it is a set of mental skills that we can develop to help us live with integrity as we pursue our professional and personal goals. It's about learning to view situations objectively, responding consciously, rather than reacting mindlessly, and using the powerful energy of our emotions as positive motivators in our lives, rather than allowing ourselves to feel helpless

133

in their grip. And like any physical training program, developing mental toughness requires conscious choice, commitment, and regular practice.

The key elements of physical fitness are strength, endurance, flexibility, agility, and balance. Proper nutrition, adequate rest, and relaxation are complementary factors that contribute significantly to physical development. There are analogous aspects to "mental toughness" (Table 7.1).

The following sections describe, in turn, the elements of mental toughness and suggest strategies for enhancing each. Just as there are myriad approaches to developing your body, there are many ways to train your mind. No single approach is appropriate or effective for all; you need to discover for yourself what works best. Mental training is a *process* that requires time, patience, and persistence; it involves changing the way you think, just as making a commitment to healthful eating means a change in food choices and eating habits—not short periods of dieting after which you return to your former ways. You need to begin slowly, making new choices in small matters—taking "baby steps" if you will—and gradually working up to dealing with more complex issues that trigger stronger emotions. It begins with accepting yourself as you are—right now—and taking responsibility for your own development. It requires patience, perseverance, regular practice, and self-discipline, like training for a marathon. But the rewards far outweigh the effort: improved effectiveness, greater satisfaction, and peace of mind.

Mental Strength

Physical strength refers to the ability of a muscle, or group of muscles, to contract against a resistance, that is, to produce or exert force. It is important for stabilizing the joints that the muscles cross, maintaining proper body alignment—particularly the trunk (i.e., the core muscles)—and performing physical activities. The benefits of strength training to athletic performance are obvious: all else being equal, the stronger athlete will outperform the weaker. What is less obvious, perhaps, are the benefits to general health: prevention of common postural problems (e.g., low back pain), protection of joints against injury (e.g., strains and sprains), increased satisfaction with and participation in physical activity, and, because of the increase in activity, enhanced cardiovascular conditioning and tension relief.

Mental strength, likewise, has as much to do with stability, force, and enhanced performance, and the benefits are analogous: a greater ability to stay "centered" and focused (especially during stressful times or "pinch periods"[1]), less tendency to overextend oneself, increased satisfaction with oneself and one's choices, enhanced persistence, and the ability to

---

[1] Defined in the chapter "Time Stress."

Table 7.1   Elements of Physical Fitness and Analogous Mental Fitness Skills

Physical fitness	Mental fitness
Strength The ability of a muscle or group of muscles to contract against a resistance; includes (a) core muscles and (b) other skeletal muscles	(a) Core beliefs and values: belief in your worth and worthiness as a person; the root of self-esteem and self-confidence (b) Strength of will: the ability to make choices consciously (however difficult) and take appropriate action; the root of conviction, determination, and commitment. Leads to self-control and "response-ability"
Agility Multidirectional speed; the ability to stop, react, change direction, and start again, all in a split second	Ability to stop behaving mindlessly, disengage emotionally, change your focus quickly, and return to mindful choosing and acting. It is the core skill in moderating attachment, minimizing perfectionism, and overcoming procrastination
Flexibility The degree of normal extensibility or range of motion within a joint and corresponding muscle groups	Ability to accept reality as it is, hold realistic assumptions, and think creatively about situations in order to generate alternative ways of approaching, interpreting, and (ultimately) solving problems and managing conflict; intellectual and emotional resilience
Balance The ability to maintain equilibrium when stationary or moving	Ability to stay focused and think clearly—without distortion—so as to maintain emotional equilibrium and sound judgment
Proper nutrition Providing appropriate nourishment to the body through healthful eating	Ability to nourish and support oneself through, for example, positive self-talk, affirmations, positive associations and environments, positive choices
Adequate rest and relaxation	Ability to relieve tension and renew emotional and physical energy
Endurance One's relative ability to continue exercising at a given rate or intensity	The patient, persistent application of all mental toughness skills; enhanced when individual abilities are well developed (and vice versa)

moderate emotion and put emotion to effective use. It has two complementary aspects: "core strength" and "strength of will."

"Core" Mental Strength

Core mental strength is a function of our core beliefs and values and is characterized by strong self-esteem and the self-confidence, determination, commitment, and the conviction that it engenders. We all hold beliefs that affect how we view ourselves and the world, how we act and interact, and how we interpret our experiences. When these beliefs are positive and self-affirming, we are more likely to approach life with confidence and realistic optimism and respond to conflict with a certain amount of detachment and objectivity. Our self-esteem is strong and we are able to function effectively.

But when these beliefs are negative and self-critical, we are vulnerable to feelings of insecurity, doubt, and indecision. When faced with challenges or conflict, we tend to lose confidence, take things too personally, engage in negative self-reflection, and end up emotionally exhausted. Our self-esteem suffers and, with it, our confidence and effectiveness.

Developing and maintaining strong, core beliefs, therefore, are essential to mental toughness. In fact, it is the foundation. Just as strong "core" muscles (abdominals and back) provide a powerful base for all athletic performance—regardless of the sport—so, too, core mental strength provides stability and a sense of personal conviction.

Strength of Will

Mental strength involves a second, complementary aspect, "strength of will": the ability of the mind to control thought and action; that is, our ability to make choices consciously (however difficult), harness our emotional energy, and use it to serve us by taking appropriate action. It is the root of determination, commitment, and self-control. Just as strong skeletal muscles are the foundation of a powerful body, strength of will is the power behind personal action.

When we live in ways that are consistent with our inner values and beliefs and respond to situations and people accordingly (i.e., "living with integrity"), we minimize mindless reactivity and the sense of helplessness that it engenders. But this is no simple task. It is far easier to react than to pause and think, to feel helpless in the face of conflict and stress than to take action in areas where we do have control. It requires a willingness to accept and take responsibility for our thoughts, feelings, and actions and for the practical and emotional consequences of our choices.

Develop "Core Strength"

The culture of science is competitive and often adversarial. We compete for positions, research funding, the best graduate students, recognition... We often experience more "failures" than successes. This is the same for our research activities. With strong core mental strength, we will not take setbacks personally but respond to them in ways that are productive and effective.

Affirm Yourself

Are your basic assumptions about yourself affirming and encouraging or negative and self-critical? These beliefs will affect how you behave and how you interpret your experiences; they may, in effect, become self-fulfilling prophecies. To develop any sense of inner strength, therefore, it is important that you believe in your inherent worth as a person, have faith in your abilities, and trust your judgment. This may be uncomfortable at first. It may feel as though you are puffing yourself up with pride. But you will not be. This is not to suggest that you cultivate an inflated view of yourself and your place in the world or hold unrealistic beliefs about your intelligence or abilities, rather, that you choose to believe in the basic premise that you have inherent worth as a human being and to have a gentle confidence in yourself based on sober judgment.

Treat yourself with dignity and respect—and expect others to do the same. Give yourself the benefit of doubt. We all make mistakes, but too often we may fall into the habit of self-blame and negative self-reflection. This does nothing but undermine our confidence and self-esteem. Yes, it is important and necessary to be disciplined and firm with ourselves—when appropriate and within reason—but beyond that, we need to be gentle, especially when we are not meeting our own professional or personal goals. Forgive yourself, affirm yourself, learn from your mistakes, and move on. And if others treat you inappropriately, give them the benefit of doubt too (e.g., perhaps they have jumped to conclusions), forgive their prejudices, and gently insist on being treated with professionalism. Respond with assertiveness, not aggression.

Engage in Positive Self-talk

Just as athletes strengthen their core muscles through a regular regimen of specific exercises, you can enhance your self-esteem and self-confidence through positive self-talk. This is a particularly effective strategy for balancing the potentially negative effects of working in a highly competitive field—and the inevitable experiences of criticism, failure, disappointment, etc., that come with it. When practiced consistently, it can become a natural part of your thinking so that, during times of stress, you will automatically use it to support yourself and cope more effectively. In effect, you will be programming your mind for optimal mental fitness and performance.

Positive self-talk—or "positive affirmations"—involves making statements to yourself that are encouraging, personal, and specific to the situation. These are usually short sentences expressed in the active (rather than passive) voice. Obviously, the topic will vary with the need. The key is that what you say has meaning for you and resonates with your experience. Examples include:

I am a worthy person, intelligent, capable, and competent.

I can make a difference in the world.

I can learn the skills that I need to be more effective.

I can develop new ways of thinking about and interpreting situations and experiences that will affirm and support who I am rather than undermine my sense of "self."

One method of learning (and mastering) the technique is to link it to physical practices, such as yoga or tai chi. Repeat the phrases regularly and mindfully (like mantras) as you move from one position to the next. In time, they will become a natural part of your thinking.

Return to Your Core Values

Each of us holds core values and beliefs that inform who we are and what we consider to be important in life. When we have a clear sense of this, our priorities will become obvious. We will know why we've chosen a life in science and what we want to achieve in our professional and personal lives. By reminding yourself of these values (especially during pinch periods) and making choices that are consistent with them, you will enhance—rather than compromise—your core mental strength. (The chapters "Career Management," "Time Stress," "Transitions," "Climbing the Ladder," and "Balancing Professional and Personal Life" all refer to the importance of understanding and honoring your core values.)

Review Your Successes to Date

What better way to affirm your abilities than to review your successes to date? Everything you've accomplished—professionally and personally—

proves that you are capable, that you can make a difference and have. Your accomplishments are facts of history that cannot be refuted. Period. Regardless of what you are involved in now, or what the future holds, you have been successful.

Gaining admission to graduate school and earning your Ph.D., for example, are accomplishments well worth celebrating. The selection criteria were rigorous and the process, demanding. You were chosen from a pool of highly qualified candidates, fulfilled all the requirements of the program, and proved yourself fit.

You are competing against your peers for research funding, publications, and positions. They simply are not awarded without thorough scrutiny by external evaluators. So when you succeed, you can be sure that others with greater knowledge and experience have judged you to be worthy.

Affirm your every accomplishment and enjoy them with satisfaction—even with modest pride—what you have achieved. And do not limit yourself to your professional activities. Any recognition from others, however modest, demonstrates that you have distinguished yourself in some way. Earning teaching awards, gaining recognition for mentoring activities, being interviewed by local media, and receiving invitations to speak to community groups, for example, are all worth celebrating.

Remember, too, that achieving your own predetermined goals is also cause for celebration—perhaps more so because the motivation and rewards come from within. For example, repaying your student loan, providing needed support to family members, taking a leadership role in your community of faith, competing in your first 10-k race, earning a role in an amateur theatrical production are accomplishments worth remembering.

An excellent way to review your professional achievements is to update your résumé or Curriculum Vitae. This needs to be done on a regular basis anyway (and added to your personnel file so that your academic institution or employer has an up-to-date record); why not make it part of your self-affirmation routine? The simple act of adding activities and accomplishments to this written record will be encouraging because you will see what you have done and how far you have progressed.

Keeping tangible reminders of your successes—and referring to them when needed—is another very effective strategy for enhancing your belief in yourself and sense of accomplishment. You may choose to create a "positive feedback" file, for example, in which you bring together letters of support; appointments and promotions; thank you letters from students, community organizations, and the like; copies of certificates and awards; notices of funding. These are testaments to your continuing success. But remember to review the file periodically and take the praise to heart; a constant "diet" of positive affirmation will go a long way to strengthening your self-esteem and raising your self-confidence.

Seek Inspiration and Encouragement from Others

An extension of creating a "positive feedback" file is actively seeking inspiration, encouragement, and support from others. Reading works that affirm your beliefs, values, and/or goals can offer tremendous reinforcement and motivation. Examples include writings by and about exemplary people; favorite passages of literature, poetry, and music (lyrics); inspiring religious texts; uplifting novels; and works that enhance learning and insight. Some people keep a written copy of favorite inspirational quotes in their offices and read them each morning.

Remember the accomplishments of scientists who have gone before you. Remind yourself that your work is building on that of the "Greats." Use them as role models, as "historical mentors." Ask yourself how they would have viewed themselves and how they supported themselves during times of stress. Draw encouragement from the progress that has been made in the culture of science since their time (e.g., increasing participation rates of women, access for underrepresented populations, numbers of job opportunities, levels of funding). By comparison, your struggles may seem less onerous.

Ask trusted friends, colleagues, and mentors for support; they believe in you and want you to succeed. They can become "your day-to-day sustenance," and during times when your confidence is flagging, they are more objective than you and can offer new perspectives that may help you to cope more effectively. Trust their support. Trust in their confidence in you.

Developing and enhancing core strength are an ongoing process. With practice, healthier mental habits will result and you will become more proficient at dealing with conflict and stress. You will be less inclined to question your worthiness and abilities and more inclined to focus your mental energies and emotion *outward*, toward action and problem solving, rather than inward.

Develop "Strength of Will"

Until one is committed, there is hesitancy, the chance to draw back, always ineffectiveness, concerning all acts of initiative (and creation).
William H. Murray, Mountaineer, in
The Scottish Himalayan Expedition (1951)

An excellent way to enhance your strength of will is to develop the *habit* of making thoughtful choices and following through consistently. This is achieved through persistent practice. Developing a new habit takes time and patience and will seem, at first, to require great effort. But as you continue to make mindful choices, you will begin to experience the benefits.

Your motivation will increase. And, eventually, the new behavior will become a part of who you are.

This phenomenon is similar to beginning a new fitness regimen: for the first six weeks, exercising regularly and eating properly require a daily recommitment to the goal. But at the end of that period, your fitness level has increased significantly for you to feel the difference: a stronger body and cardiovascular system, greater flexibility and resistance to stress, better fitting clothes, and a more positive attitude toward life. Not only that, what began as a chore and perhaps even a struggle has become a normal part of your lifestyle that you look forward to and enjoy. Again, it is important to begin with small changes. Be patient with yourself and accept that any change in behavior takes time.

It is important, when developing this new habit, not to undermine your efforts by allowing yourself to question your decisions. Do not look back or second-guess yourself. Consciously choose not to entertain doubt. Once you have examined the issues as carefully as you can, considered all aspects to the best of your ability, sought more information as appropriate and incorporated it into your analysis, and made your choice, press on boldly, with confidence and conviction (even if you don't yet feel it). You will thus be able to focus all your mental resources on taking action and thereby make your choice succeed.

Remember: when you make decisions this way, you are making them with the best information available at the time. You will never have all the data. The point is to make the decision and move on. If and when something new comes to light, you can revisit the topic and make a new decision, if appropriate.

As with developing core strength, you may find it helpful to use positive self-talk to affirm and support yourself. For example, repeating the following phrases when you are tempted to doubt yourself may provide encouragement and strengthen your resolve:

I have confidence in my ability to analyze the situation objectively and decide accordingly.

I am making the best choice possible, given the circumstances.

In making this decision, I am being true to my values and beliefs.

Other statements that may provide motivation are:

The mentors whom I admire have been in similar situations and have maintained their professionalism.

Sometimes we have to choose between bad and worse. I am making the best decision in a difficult situation.

I will press on, regardless.

And we must not overlook the important role of *imagination* in motivation. Goethe, the German poet, dramatist, novelist, and scientist, expresses it best in his insightful quote:

> *Whatever you can do or dream you can, begin it. Boldness has genius, power and magic in it. Begin it now.*

MENTAL AGILITY

> *Nothing in life is to be feared. It is only to be understood.*
> Marie Curie, Nobel Prize Laureate in Physics (1903) and Chemistry (1911)

A critical element of mental toughness is mental agility: the ability to stop behaving mindlessly, to disengage emotionally, change your focus quickly, and return to mindful choosing and acting. It comes into play during the *initial* stages of any stressful situation (e.g., conflict, "pinch period") when there is a potential to become carried away by a flood of feeling, and during periods of *sustained effort* (e.g., when writing grant proposals, meeting tight project deadlines), when you are vulnerable to exhaustion. Mental agility is the core skill in moderating attachment, minimizing perfectionism, and overcoming procrastination. Without it, precious mental and emotional resources are wasted, focus is lost, and effectiveness is compromised.

Its importance is analogous to that of "multidirectional speed" in competitive sports such as soccer (football): when advancing the ball toward the opponents' net, an offensive player must slow down, stop, change direction, and start again in a split second, as she anticipates and responds to the actions of the defending team. She may choose to carry the ball around the fullback herself, or let go of it by passing to an unguarded teammate. Either way, the ball remains under the control of the offensive team. And during these maneuvers, the player's focus is on scoring a goal—not on the challenges posed by members of the opposing team.

So, too, in science: We must be able to quickly and effectively deal with problems and distractions, so as to stay focused on advancing our professional and personal goals. This involves becoming aware and learning to stop, let go, and move on.

Develop Awareness

The first step in developing mental agility is learning to recognize the internal signs and symptoms associated with states of heightened arousal. We've all experienced them: the muscular tension in our throats, chests, and shoulders; the knot in our stomachs; the shallow breathing (or holding of our breath); the narrowing of our focus and attention; even, perhaps, a welling up of tears or a desire to lash out in anger. All these are a normal part of the human "fight or flight" reaction and can be triggered by myriad circumstances (e.g., anticipation of an upcoming performance review, feeling overwhelmed by the mounting pressures of multiple work responsibilities, conflict with a colleague, extended or excessive concentration on a crucial task caused by perfectionism, missing an important deadline because of procrastination).

All too often, our own thoughts can compound the problem. If we do not view the triggering situations clearly or if our thinking is distorted (see "Mental Balance" section), we may well react even more strongly. Being able to distinguish the situations that engender strong emotional reactions and becoming aware of our thoughts before, during, and after these "crises" are crucial to defusing our reactions and learning to act responsively under pressure.

What is important is to recognize them quickly and deal with them immediately. The goal is to prevent our normal heightened arousal from escalating to the point where we feel helplessly caught up in a flood of feeling that controls our actions—that is, where we are "in the grip" of emotional reactivity. We want to respond in ways that will enable us to maintain our professionalism, rather than react in ways that may undermine the respect of our colleagues or make us feel bad about ourselves in the long run. Understand that these reactions are *normal* and may take some time to be able to recognize. But you will learn. Awareness and acceptance of this are also part of the process.

Stop the Escalation of Tension

Many techniques exist for moderating the effects of mounting *physical* tension. Consciously relaxing your muscles; taking long, slow, deep breaths; looking away; turning your body away; and/or stepping back slightly from the person or object that is the catalyst of the reaction—all these strategies will help in a crisis.

Stopping yourself from becoming carried away *emotionally* requires commitment, discipline, and patient persistence (and a healthy sense of humor about yourself), for it requires changing what you think and say to yourself; it requires a change of *mind*. You need to consciously choose not to let your normal emotional reactions *dictate* your responses. By defusing your automatic reactions, you will be able to choose how you will respond.

One approach is to tell yourself to stop. As soon as you become aware that you are reacting to a situation, repeat to yourself—firmly, but kindly—a word or short phrase that is personally meaningful and effective in grabbing your attention, for example,

No. No.

Stop. Stop.

Pause. Pause.

Wait. Wait.

Gently. Gently.

Step back. Step back.

Push away. Push away.

Make it imperative. Do not reason with yourself or give yourself the option of not obeying. Regardless of whether you say it silently or aloud, the aim is to stop the reaction from escalating. For greater effect, link your stopping commands to the tension-reducing physical behaviors already discussed. For example,

Triggering event: Stepping up to the podium to deliver an important seminar.

Stopping behaviors: Inhale deeply as you approach the podium, exhale slowly, and repeat to yourself, "Breathe. Breathe. I am well prepared, capable, and confident."

Triggering event: Confrontation with team member or supervisor.

Stopping behaviors: Turn one shoulder away from the person (so you no longer are facing them squarely) and say to yourself, "Gently. Gently. I will listen to what they're saying, not how they're saying it."

Triggering event: Sitting for hours at a computer to meet a crucial deadline, you become aware of your stiff neck, shoulders, and lower back.

Stopping behaviors: Look away from the computer (to some point at a distance), stand up slowly, and stretch your body as you say, "Relax. Relax. I'm making excellent progress."

Let Go So You Can Act Mindfully

The advice that I give to graduate students who have had a bad experience is this: "Deal with the situation as objectively as possible, learn what you can from it, then 'press the reset button.' Holding onto bad feelings serves no useful purpose and will hurt you more than anyone else."

Melissa Franklin, Physics, Harvard University

It is not sufficient simply to *stop* reacting to situations and people; we need to *let go* in order to move on to thinking and acting *mindfully*. By learning how to *consciously* release your mind from a narrow, single-minded focus, you will avoid becoming "stuck" in a cycle of reactivity. This is not to suggest that concentrated attention is undesirable; on the contrary, it is extremely adaptive—essential, even—in coping with and fulfilling the multiple responsibilities of our personal and professional lives. The critical point is to recognize when it is becoming detrimental and to let go as quickly as possible, so that you can move on to more effective and adaptive ways of thinking and acting. The benefit of "disengaging" emotionally is that it enables you to examine the situation more objectively and gives you a sense of control. More importantly, it minimizes any tendency to internalize and personalize the situation. And, by re-engaging with a clear direction and purpose (acting mindfully), you will use the "emotional momentum" of your reactions to your *advantage*.

"Letting go" involves consciously redirecting your mental energy toward positive ends. Obviously, this requires clearly defined goals (based on your values, priorities, etc.), an understanding of the choices that would best advance them (aspects of "mental flexibility" that are explored in the next section), a commitment to taking the appropriate action, and the will to do so (i.e., mental strength). Though the specifics will differ with the individual and the circumstances, targeted, affirming self-talk is the key to letting go and moving on. When practiced consistently, you will develop confidence in your self-control and in your ability to handle any situation with professionalism.

Take Your Emotional Reactions Less Seriously

A useful first step in letting go is to give your emotional reactions less weight in your assessment of a situation or issue, that is, to take them less seriously. Just because you react strongly to something does not mean that whatever triggered the reaction is insurmountable. For example, a critical experiment has been giving you problems. You've run it four times and still it's not working properly. After each attempt, you analyzed what may have gone wrong and made adjustments. This time, you *know* it will work. With tense expectation, you open the incubator door. Remove the plate. And look...NOTHING. No colonies. NOTHING! You react.

Of course this reaction is natural. And that's the point. Instead of "fighting" the reaction, accept it for what it is: an *expression* of disappointment, frustration, perhaps anger, just as laughter is one expression of amusement. Try humorous self-talk to encourage yourself to take the reaction less seriously. Or gently make fun of yourself to help you let it go:

I could feel that reaction coming. But hey! This time, it's only a "6" out of "10" on the intensity scale.

Another approach is to act "as if" you feel calm and in control. This technique is discussed later in this chapter, under "Maintaining Emotional Equilibrium" in the section on Mental Balance.

Let Go of "Shoulds"

A common trigger of emotional tension is the word "should": *"I should have started writing this grant application sooner; I'll never get it done!" "My supervisor should take more interest in my progress." "I should have earned that promotion; the system should be fair!"* These judgments—of ourselves, of others, of the world—are based on expectations that are not consistent with what we are experiencing. However reasonable the expectation or accurate the judgment may be, when we focus on what "should" be rather than on accepting *what is*, we risk becoming fixed on the thought and immobilized by the associated negative feelings.

One way to avoid this is to replace the word "should" with the phrase: *"it could..."* or *"it would be better..."* immediately you become aware that your thinking involves "shoulds." By doing so, you will be more able to let go of the expectation and/or judgment and minimize the possibility that you will become "stuck" in a line of thinking that will only lead to reactivity. For example,

Triggering thought: "I feel my colleagues are ignoring my contributions to this discussion. They should recognize and give me credit for my input!"

Defusing thought: "I feel my colleagues are ignoring my contributions to this discussion. It would be better if they recognized and gave me credit for my input; I have much to offer. Perhaps I'll speak with <a trusted colleague> about this later; he or she may have some suggestions for me."

In this example, the modified phrase helps the thinker to detach herself from reacting emotionally to the situation. And the idea to speak privately with someone afterward enables her to redirect the energy associated with her frustration toward a positive future action.

Certainly, there will be times when it is quite appropriate to give considerable thought to what "should be" (e.g., in order to improve the situation and/or environment for yourself and others). But the time is *not* when you're already in a heightened state of arousal. Once you've established some emotional distance, you'll be able to revisit the issue and decide whether it truly is something worthy of your effort and for which you have time and energy. Unfortunately, there will be many occasions when we have to leave important battles to be fought at a later date or by others. If addressing the issue is a priority, then of course, do something about it. If not, make a conscious decision to let it go and move on to more important things. When the issue comes up again (which it will), remind yourself of your priorities, the choices you've made, and choose again to let go of the issue.

Any expectation that we may have that others will: agree with us, affirm us, acknowledge that we are "right" or acknowledge when they are "wrong," cooperate, be reasonable, logical, and consistent, or behave with civility, can lead to thinking that involves "shoulds." The best way to deal with triggers such as these to is identify them and develop phrases to defuse them. The following examples may give you some ideas:

Triggering thought: "Dr. X had nothing positive to say about my presentation; his criticisms went on and on. I should have practiced more and been better prepared."

Defusing thought: "I've considered his criticisms and he did have some valid points. But he also had a lot of trivial complaints that had little to do with science or my reasoning. Clearly, he was 'puffing off'."

Triggering thought: "I know I'm right on this point, but she just won't admit that she has made an error in her interpretation. She should be professional enough to admit that she is wrong."

Defusing thought: "She has a right to be mistaken."

Sometimes, choosing to "forgive" the person is the best way to let go of your expectations of them. This is especially true when letting go of our expectations of ourselves.

Triggering thought: "The personal remarks she made were completely inappropriate and very hurtful. She shouldn't treat me like this, especially in front of everyone."

Defusing thought: "I choose to ignore her insensitive remarks and refuse to take them personally. I forgive her for her lack of sensitivity."

Triggering thought: "I've a report to write, an overdue grant to review, a lecture to prepare for tomorrow, and now, my son is running a fever! I should be more organized. I should have started earlier. I should have anticipated this."

Defusing thought: "I am doing the best I can under the circumstances."

Actively Ignore

While remembering some events in our past can help us deal with our present circumstances (e.g., deriving confidence and encouragement from having successfully solved an earlier problem in experimental design, being awarded a research grant or having earned a promotion), dwelling on others can be counterproductive. Grievances, hurts, guilt, embarrassment, frustration, anger...these emotions and others can be re-experienced if we continue to relive the situations that engendered them. When our minds are attached to such preoccupations, we will waste precious mental energy and time and quite possibly lose confidence in ourselves and our trust of others.

Certainly, it is important to reflect on one's experiences in order to learn from them (mental flexibility). But having done so, we need to let go of any associated negative emotions so that circumstances in our current situation will *not* trigger the emotional reaction associated with past experiences. By learning to recognize these stimuli and developing self-talk to help us ignore them, we will be able to avoid mindless reactivity.

Just as athletes must let go of their losses, forgive their errors, learn from their mistakes, and focus completely on their present task to achieve peak performance, so too must we moderate our attachment to unhelpful, rigid, or "obsessive" thinking and other negative mental states. When we can actively ignore such thoughts, we will be able to put the full power of our minds to positive productive ends. The following are but a few examples of helpful thinking; develop phrases that speak directly to your needs and experiences.

I've been in this situation before; I choose not to let this bother me.

This issue is not important enough to waste my valuable time. I will let it go.

She's trying to bait me. I won't give her the satisfaction of a reaction.

I am not a victim. These strong feelings will subside.

"Declutter" Your Mind

Another way to let go of triggering thoughts is to view them as unnecessary "clutter" in your mind. Whenever you become aware that you are dwelling on the past or worrying about the present or future, ask yourself whether the activity is energizing.

When I think about <fill in the blank>, does it lift my energy?

If not, visualize yourself sweeping the thoughts aside (or flushing them down the toilet) and focus on something positive.

Overcome Procrastination; Minimize Perfectionism

Procrastination and perfectionism are like two sides of the same coin. While one is characterized by not beginning early enough (procrastination), the hallmark of the other (perfectionism) is not stopping soon enough. Both are very common (but unhelpful) strategies for decreasing the stress associated with work and often compound each other: many people take too long to get going and, once moving, take the work too far. Just as athletes who do not begin training early enough—*and* those who *over*train—will not achieve peak performance during competition, so too will procrastination and perfectionism compromise the effectiveness of a scientist.

The factors contributing to the development of either habit are many and varied. They include the competing and changing demands of busy professional and personal lives; expectations, circumstances, and unpredictable events over which we have no control (e.g., poorly articulated performance standards, the last-minute assignment of a critical progress report by a supervisor); underdeveloped activity planning and scheduling skills; individual differences (e.g., in personality, values, goals, definitions of "success"); and lack of confidence in our abilities. While some factors are under our control, many are not. The point is to quickly recognize those that we can influence (develop awareness), do something about them (act mindfully), and let go of the rest. Well-developed mental agility will enable you to do just this.

How we make choices regarding the many priorities competing for our time and attention and organize ourselves to accomplish our work are not matters of mental agility. The chapters "Career Management," "Time Stress," and "Balancing Professional and Personal Life" address strategies that can help.

Lack of self-confidence and the related fear of not being able to meet expectations of your performance, on the other hand, *are* factors that can be moderated by mental agility. They inspire a mindless *focus on self* rather than on *action that will resolve* the task at hand and are powerful triggers to procrastination and perfectionism.

From time to time, we all experience feelings of uncertainty about our abilities and concern about our performance, especially when facing new situations. This is normal and, to a certain extent, quite adaptive: it motivates personal and professional development and inspires achievement. After all, being *too sure* of ourselves can lead to complacency, lack of preparation, and poor performance, just as elite athletes who enter a match feeling overconfident can be beaten by poorer players. The point is not to eliminate all uncertainty—which is impossible—but to recognize when it becomes a trigger to maladaptive behaviors and to redirect your thoughts and energy toward positive action.

By identifying the specific thoughts that lead to feelings of insecurity or fear, you will be able to develop positive self-talk that will counteract them and defuse mindless reactivity. And by identifying alternative actions that you can take, you will be able to direct your energies toward more adaptive behaviors.

For example, for those prone to *procrastination*, the point of mental agility is to inspire you to *begin* sooner than you normally would, before you feel "ready." So the phrases that you develop need to be aimed at decreasing the anxiety related to early starts.[1]

Triggering thought (lack of self-confidence): "For this new project I must know the literature thoroughly, talk with all my colleagues to understand their diverse opinions on the topic, and think through every possible experiment. I can't begin until I do."

Defusing thought (lack of self-confidence): "I have used due diligence to understand the background for this project, identified the question that I'd like to pursue, and determined how to proceed. Now is the time to begin my experiments. Spending more time thinking will not generate data. And who knows: the initial results may suggest a different—and better—direction that I can't possibly anticipate now. Spending extra time to continue thinking things through is a waste of time and energy."

Triggering thought (fear of failure): "I simply must find funding for my research! This granting agency is the best choice for my field, but preparing the application is going to be a demanding task. And the success rate is so low! No matter what I do, I may not even get funded. I need large chunks of uninterrupted time to work on this so I can concentrate properly, but I only have 45 minutes right now. I can't do anything in that time."

Defusing thought (fear of failure): "I know the literature in my field, have carefully done the preliminary experiments, and have some very interesting data. It's an intriguing line of inquiry that I'd like to pursue. And it's well within the mandate of the granting agency. I'll make a start on the grant application right now, so that I have something to build on when I have a half an hour to spare. I'll be making progress and will feel

---

[1] For an excellent discussion of how to deal effectively with procrastination and perfectionism, see Boice (2000).

less anxious. And there'll be more time to show it to my colleagues and revise it, based on their feedback."

On the other hand, for those prone to perfectionism, the point of mental agility is to inspire you to stop sooner than you normally would. Chances are your work is "good enough"; it's just that you don't believe it is. The phrases that you develop need to be aimed at decreasing the anxiety related to letting go. For example:

Triggering thought (lack of self-confidence): "I'm feeling very uncomfortable with these experimental results. I should repeat the experiment again and again until I'm confident."

Defusing thought (lack of self-confidence): "It is reasonable, natural, and scientifically healthy for a scientist to question her results. I've rerun this experiment several times, using different approaches, and the results are the same. This is sufficient for this type of project. It's time to move forward."

Triggering thought (fear of failure): "I'm very nervous that the grant review panel will identify a flaw in my grant application and will reject it out of hand. I simply must have funding or else I can't do my research."

Defusing thought (fear of failure): "I've done everything that I can to write the best grant application possible. Senior colleagues who have a good track record of securing funding have vetted it and they have no more suggestions to make. This is the best I can do under the circumstances and is all that anyone can expect."

It generally is a good idea to seek the feedback of knowledgeable colleagues and/or friends (as the situation dictates) because they bring experience, objectivity, and a fresh perspective to the situation. And doing so sooner—rather than later—will give them more time to reflect and respond and will save *you* time and energy because you will be letting go sooner and moving on.

You may not experience your own procrastination and/or perfectionism as a "problem." Perhaps you are energized by last-minute deadlines and accept that you'll be working all night. Or you are willing to spend the extra time and energy to do a "perfect" job when an "average" job is all that is required. But it is essential to understand that you are not working and living in isolation from others. Your own procrastination or perfectionism can have very negative consequences for your colleagues and coworkers—especially on those who report to you—as well as for your friends and family.

And be warned: There always will be occasions when you will have to cope with the consequences of *someone else's* procrastination and/or perfectionism. Their behavior is not under your control (however much you may try to influence it), so it is unlikely to change. You will have to accept the reality of your situation and minimize its impact on you. Applying the strategies of mental agility (especially "letting go") will help you to defuse the frustration and/or anger associated with the negative effects (on you) of their poor choices and help you to redirect your thoughts and energies toward completing the task at hand.

Being able to stop behaving mindlessly, disengage emotionally, change your focus quickly, and return to mindful choosing and acting all are important aspects of "mental agility." With patient practice, your skill will develop and you will become more adept at moderating attachment, minimizing perfectionism, and overcoming procrastination. Implicit in replacing "triggering" thoughts with those that defuse and redirect your focus and intention is the ability to think creatively about your situation and circumstances and generate more helpful alternatives. It requires "mental flexibility."

MENTAL FLEXIBILITY

Minds are like parachutes. They only function when they are open.
                        Sir James Dewar, Physicist and Chemist

The human body is designed for action. Muscles, tendons, and ligaments surround the joints, provide support and stability, and control the movement of the bones. The degree of normal extensibility within a joint and the corresponding muscle groups is termed "flexibility." If the muscles are too tight, range of motion decreases; if too loose, the joint loses stability. Both will compromise athletic performance and can result in serious injury (e.g., a torn muscle, if flexibility is poor, or dislocated bone, if the joint is unstable). Strength and flexibility are important to mobility, and athletes develop both, to the degree required for optimal performance in their sport.

Similarly, mental strength with little mental flexibility is not an effective combination. By developing mental flexibility, we are more able to accept the world as it is, form reasonable assumptions, set realistic expectations (of ourselves, of others, and of life), and think creatively about situations in order to generate alternative ways of approaching, interpreting, and (ultimately) solving problems and managing conflict. It will expand our options, inform our actions, and increase our intellectual and emotional resilience.

Develop Acceptance

It may be necessary temporarily to accept a lesser evil, but one must never label a necessary evil as good.
Margaret Mead, Anthropologist

A key component of mental flexibility is the ability to acknowledge and accept the world as it *is*, with all its imperfections, and to see ourselves and others clearly, without judgment. Acceptance does *not* mean that we agree with, support, or are satisfied with reality, nor does it imply that we believe that everything is perfect. Rather, it is the ability to begin any analysis or reflection from a solid understanding of what "is," so that our assumptions, expectations, and questions are reasonable (something scientists are trained to do in their research).

There are many things over which we have little or no control that can be a source of great irritation, frustration, or anger. For example, we have spent the past five years developing a new pharmaceutical that we believe is quite promising, when the project is terminated because the priorities of the company have changed, or we discover that the analysis of our data is faulty because the values that we included from our collaborators had not been converted to the same units that we were using. Dwelling on situations such as these can trigger counterproductive thinking and the development of strong negative emotions. By choosing to let go of such thoughts as soon as we're aware of them, we will avoid falling into mindless reactivity. We will be free to focus our minds and emotions on what we choose. Phrases such as the following may help:

This is the way it is right now. I don't like it. But I choose to accept it.

I have no control over the behavior of others. Nor can I control what they think. I choose to let go of my expectations of them.

I will accept. Let go. And move on.

Women, in particular, often are very hard on themselves—a tendency that can add enormous stress and insecurity. It is important to accept that it will take time and patience to become aware of escalating tension, of emotional triggers, and learn how to defuse them. Be reasonable with yourself; a little time and emotional distance will help you gain perspective, understanding, and ideas for what to do differently. And remember that it is a *process*. Accept that the process requires time. Again, be gentle with yourself as you patiently and persistently develop new ways of thinking.

Think Creatively to Solve Problems and Manage Conflict

The second aspect of mental flexibility is the ability to think creatively about yourself and your circumstances in order to generate alternative ways of approaching, interpreting, and (ultimately) solving problems and managing conflict. By learning to examine situations from different perspectives and choosing the best option(s), you will minimize your tendency to become "stuck" in mindless reactivity and will, instead, move forward in focused action. The following are just six strategies for reframing potentially triggering situations. Many more exist. Explore others and discover what works best for you.

Let Your First Assumption Be Positive

One strategy for reframing potentially triggering situations is to make your first assumption a positive one, that is, to give yourself and others the benefit of the doubt. By doing so, you will be more likely, willing, and able to continue listening with acceptance (i.e., without becoming defensive), to understand more fully, and to reach a positive (or, at the very least, neutral) resolution. If, for example, in your first job performance review, your supervisor commends you for the quality of your bench work, but questions the length of time it takes you to do the experiments, do not assume immediately that you are not working efficiently or that he or she is criticizing you. (Jumping to conclusions like this is an example of a "distortion" in thinking, discussed in the next section.) Assume first that the time you took was what was required and that your supervisor simply wants a better understanding of the protocols.

If someone clearly is offering criticism, choose to view it as feedback on your performance at a specific point in time, and not a prediction of your future performance and potential for success. On no account is it wise to consider any criticism as a negative statement about your intelligence or worth as a person. Determine whether the criticism has any merit. If there is some truth to it, reflect on the situation and calmly decide what you can and will do about it. If not, choose carefully how you will respond. You may, for example, ignore it as an inappropriate outburst (e.g., with the thought *"There he goes, pontificating again!"*), write a rebuttal (e.g., of manuscript criticisms), or speak privately with the person after first considering what you wish to say.

Similarly, if a colleague makes a comment to you that is—from any perspective—inappropriate, assume first that they're speaking from ignorance or are under tremendous stress, rather than out of deliberate maliciousness. Even if your positive assumption proves false, you'll have maintained mental balance (see next section), so you'll be able to respond professionally.

Determine the Facts

Another strategy is to apply the scientific approach, that is, to gather facts through observation and inquiry (e.g., ask probing questions) and to evaluate the situation as dispassionately and as critically as possible. Look at the situation in context and from a variety of perspectives. Perhaps, for example, your project was not funded because your proposal did not adequately considered the priorities of the granting agency and what research they usually support. Can it be rewritten to emphasize *their* interests and to make clear the value—to them—of the work that you wish to do? Can you identify other groups that may be more interested in what you want to do and more likely to provide funding?

If, on the other hand, it's a matter of interpersonal conflict, consider everyone's perspective and ways of interacting. What are the goals of the specific parties engaged in the problem? What have you learned about them from your previous encounters? What experiences have others had? (Rely on your support network to help you gain perspective on this.) What have you observed about how they relate to others? Do you respect them? Do others respect them? Is this an unusual occurrence? Do they behave this way with others? Are your goals, values, or priorities in conflict?

Ask the "Right" Questions

As scientists, we are trained to ask probing questions and to approach our work with a critical eye and a healthy amount of skepticism. These skills are essential for creative problem solving and are an inherent aspect of "mental flexibility." But sometimes we become so attached to the questions we pose that we lose our ability to see better alternatives. If this results in mindless reactivity, our effectiveness is diminished.

There are many questions that we can ask about our work, our relationships, and ourselves that will influence the fulfillment of our goals and needs and enhance or diminish our confidence and self-esteem. Dr. Melissa Franklin, the first female professor of physics at Harvard University, has a useful way of viewing questions that many of her graduate students find helpful. She describes them as belonging to one of three groups, depending on their effect on the questioner:

"Interesting" questions inspire professional and/or personal growth, development, and advancement. They keep us focused on matters over

which we have some influence or control and encourage us to take positive action. Examples include "What aspect of my current research so engages me that time passes without my being aware of it?" "How can I use the reviewers' comments to help me strengthen my grant proposal so that when I resubmit it, I will have a higher probability of being funded?" "How can I prepare for the next meeting with my colleague so that his rudeness won't bother me so much that it interferes with our achieving what needs to be done?"

"Uninteresting" questions, however intriguing or emotionally engaging, consume precious time and mental energy with no net benefit, for example, "Why didn't my supervisor assign this to me earlier? She always seems to do things at the last minute!" and "Everyone seems so much more knowledgeable than I. Am I smart enough?" These lines of inquiry do not inspire action. In fact, dwelling on them too much can become "dangerous."

"Dangerous" questions are precisely those that can inspire attachment to thoughts that trigger mindless reactivity. They have damaging effects on our productivity and can easily undermine our confidence and self-esteem. Examples include "I have good ideas! Why are others receiving credit for them and not me?" and "This system isn't fair! I've done all the 'right' things and still I'm not being taken seriously. Why must I be twice as good as everyone else?" These kinds of questions (however accurate the underlying judgments may be) will generate strong emotional reactions that quickly can become immobilizing if left unchecked. Nothing will be resolved if we are caught up in negative feelings. As Dr. Franklin so aptly puts it: "Being bitter is a disaster."

Certainly, we all have to face *difficult* questions from time to time, but we need to decide wisely which to entertain and which to let go. In addition, we need to consider *when* we'll pause for reflection. It is wiser to choose occasions when we have the time and energy to do the necessary intellectual *and emotional* work than when we're under tremendous stress.

The point of mental flexibility is to recognize and accept the reality of the situation and people involved, approach interactions with reasonable expectations, and ask questions that will enable you to move forward with confidence. For example, one could use the following to address the concerns expressed by the uninteresting and dangerous questions (above):

My supervisor procrastinates. This is a reality I cannot change. How can I organize myself/my work in order to reduce the impact when it happens again?

It's normal for people to seem knowledgeable about their area of expertise. And so they should. I wonder how people who don't have my knowledge and experience perceive me? Besides, the fact that I've advanced to this level proves that I am smart enough and have what it takes to succeed!

I suppose it's flattering (in an odd way) that others may try to take credit for my good ideas, but they're displaying intellectual dishonesty. I will not ignore this behavior all the time. But is this current case important enough for me to set right? And if so, what is the best way to proceed?

No system is perfect. And this one certainly is a case in point! My work is important to me and I want to continue. So what can I do now to support and encourage myself as I deal with the inequities that I experience? And what can I do, however modest, to improve the system for myself and for those who will come after me?

Reframe the Situation as a Scientific Experiment

Another way to think creatively about a difficult situation (and especially to depersonalize conflict) is to treat it as an opportunity to experiment in a living laboratory: analyze the situation, decide on a course of action, follow through, evaluate the results, and modify your action(s), if necessary. By applying the analytical methods of your scientific training to the problems that you encounter beyond your research, you may be able to maintain a greater degree of objectivity and gain a new perspective that will help you to generate ideas for better solutions.

View "Failure" as a Learning Experience

If I find 10,000 ways something won't work, I haven't failed. I am not discouraged, because every wrong attempt discarded is often a step forward.

Thomas Edison, Inventor

No "failure" is ever a waste of time and energy if you have learned from it. Consider every person a potential "teacher" and each experience an opportunity to clarify your values, goals, and priorities; to enhance your knowledge, insight, or understanding; and to practice and develop your professional, intellectual, interpersonal, and self-management skills (including problem solving, decision making, and mental toughness). Even if what you have learned is how to recognize earlier and avoid sooner the situations and people that waste your time, energy, and resources, you have gained an enormous advantage.

Use Humor: Look for the "Entertainment Factor"

A particularly useful approach to dealing with situations over which we have no control is through the use of humor. By specifically looking for aspects that amuse us and consistently focusing on these, we'll begin to associate amusement (rather than frustration or anger) with what once were triggers. As with taking our emotional reactions less seriously, looking for the "entertainment factor" is an effective way to accept things that we cannot change and remain patient with ourselves and with others.

MENTAL BALANCE

In physical fitness, "balance" is the ability to control one's body position in order to maintain equilibrium when stationary or in motion, that is, to keep one's center of gravity within one's base of support. Without it, athletes will not be able to execute the complex coordinated actions of their respective sports with the speed and agility necessary for peak performance. Likewise, "mental balance" is the ability to stay "centered" in one's thinking and feeling, especially under stress (e.g., during "pinch periods" or in situations involving conflict), that is, to stay focused and think clearly—without distortion—so as to maintain sound judgment and emotional equilibrium. When well developed, mental balance will enable you to effectively weigh the alternatives generated by mental flexibility (taking into account all relevant factors, including your current circumstances, long-term goals, the realities of the situation and people involved, and probable consequences), so that you can determine the optimal solutions, responses, and/or actions. Making the choice and following through require strength of will. Where mental agility enables us to quickly recognize and respond to triggers, defuse the associated tension, and redirect our focus, mental balance provides long-term, dynamic stability by eliminating distortions in thinking that are the triggers of reactivity.

Cultivate Sound Judgment

The importance of sound judgment is obvious. When we're unfocused and unclear, we may apply ourselves and our resources to projects that are ill conceived or "dead end," spread ourselves so thinly that the quality of our work suffers, or devote *too* much time to efforts that may not be recognized or highly valued by our employers or funding or promotion panels. We may hold unrealistic beliefs about ourselves, our abilities, and our prospects that could lead to inappropriate career choices, interpersonal conflict, or even failure. We all know scientists who think very highly of themselves but who are not well regarded by their peers because their research does not measure up.

Staying focused and thinking clearly to cultivate sound judgment require that:

- we invest ourselves in things that are important (as judged by our professional and/or personal values, goals, priorities, responsibilities, etc.)
- we have as much relevant information as possible
- our assumptions are correct (as discussed in the "Mental Flexibility" section)
- we ask the "right" questions (also defined in the "Mental Flexibility" section)
- our statements to ourselves (i.e., "self-talk") are sound

Cultivating sound judgment involves *integrating* the thinking skills of core mental strength, agility, and flexibility and applying them appropriately to the situations that we encounter in our daily lives. *How* they are applied is an individual matter that cannot be prescribed. Again, the process takes time, patience, practice, and self-kindness.

Maintain Emotional Equilibrium

Emotions are a powerful force in our lives that can influence what we think and how we behave. They can overwhelm and immobilize us (i.e., hold us "in the grip"), precipitate mindless reaction, or motivate us to take positive action. Staying focused and thinking clearly to maintain emotional equilibrium require that:

- our thinking is not distorted
- our statements to ourselves (i.e., "self-talk") are sound
- we focus on responding effectively (rather than reacting)
- we maintain a positive "can do" attitude

We cannot avoid feeling hurt from time to time; it is difficult not to take some things personally and not to feel angry or disappointed. But we are not helpless. We need not feel as though we are "victims" of our emotions, because the relationship between thoughts and emotions is bidirectional. Just as emotions influence what we think (and, consequently, how we behave), so, too, does what we think (and how we act) influence how we feel. We can change the direction of influence by *consciously choosing* to think and act differently. Though we may still feel the "sting" of emotion, we will think and behave reasonably and calmly. The following are some strategies that may help to maintain emotional equilibrium.

Eliminate Distortions in Thinking

The relationship between thoughts and emotions has long been investigated by researchers and clinicians. From their studies have emerged many popular works that offer strategies for thinking clearly and acting positively. One particularly helpful book (Burns, 1999) identifies what the author calls "distortions" in thinking that trigger reactivity and clearly describes a practical, efficacious approach to dealing with them that mirrors the mental toughness strategies discussed so far. These distortions can be grouped into six categories.

Distortions in Thinking That Trigger Reactivity

Applying a negative filter: focusing on a single, negative detail to such an extent that it prejudices your entire view. For example, the head of your research group provides strong criticism of a draft report you wrote. You found one comment to be particularly hurtful. In spite of the fact that you were more satisfied with the revised report and it subsequently received praise from your department head and the administration, you continue to focus on that one comment. It continues to bother you and begins to color your view of yourself and your Head.

Related distortions involve inappropriately (a) exaggerating the importance (and influence) of negatives and/or (b) minimizing the positives. For example, (a) it has been a long and challenging month of work on a key project. Members of the research team are tired and tempers are frayed. In the heat of the moment, you say something inappropriate to a colleague and immediately regret it. Even though your apology has been accepted, you continue to dwell on your momentary lapse and your sense of guilt overshadows all of your interactions. (a) You are about to present an important paper at the annual meeting of your professional organization and you're feeling anxious. As you think of all the "experts" who will likely be in the audience (and overemphasize their knowledge and expertise), you begin to minimize the importance of your own contributions. You lose sight of the fact that you are the "expert" on this particular aspect of the field, and you lose confidence in the conclusions that you've reached.

Another variant is actively rejecting positive experiences by giving them a negative "spin." For example, *"It was just luck that I was offered this position."* (A common belief in the "Impostor Syndrome." You discount the fact that you went through a rigorous selection process and were chosen as the preferred candidate.) Discounting the positives is also a symptom of all-or-nothing thinking (i.e., viewing things in extreme black or white categories). For example, in your performance review, your department head tells you she's satisfied with your performance in research and service but is concerned about some strong criticisms of your teaching that were raised by a few students in your introductory class. You pay no attention to the praise and think only *"I'm a failure as a teacher."*

Making "should" statements: making judgments about yourself and/or others that include statements involving "should," "must," or "ought." For example *"I should have started my review of the literature sooner." "This grant application system shouldn't be so convoluted." "These people should be more collaborative."*

Jumping to conclusions: immediately assuming the worst, without considering the facts, and drawing negative conclusions, regardless of the evidence. For example: (a) mind reading: *"My supervisor looks cross and hasn't spoken to me in two weeks. Why is he angry with me?"* (b) Fortune teller error: *"The Head of the research unit wants to see me... What have I done wrong?"*

Overgeneralization: extrapolating to future events the negative experience of a single event. For example, you are unsuccessful in your first attempt to earn funding from a philanthropic organization that supports research in your area and you think *"I'll never get funding from them!"* Taken to the extreme, overgeneralization can result in your labeling or mislabeling events and/or people (including yourself) using negative, critical, and emotionally charged language. For example, in the same situation as above, you say to yourself *"No one will fund me. I am a failure."*

Personalization: automatically thinking that you are at fault for some negative event, even though you had little to do with it. For example, you and your colleagues have been working on an important project and you have repeated an experiment several times without success. The protocol is tricky and you've done it carefully, but it fails again. You tell yourself: *"It's my fault that this experiment isn't working."*

Emotional reasoning: using the "evidence" of your negative feelings as the basis for your beliefs about reality. For example, you are working on an important grant application and the deadline is looming closer. You feel strongly that you won't complete it on time and believe *"I'm not going to get this grant in on time."*

Dr. Burns argues that *most* events, or thoughts, that trigger strong reactions involve some distortion in thinking. His strategy for debunking distortions in thinking and establishing emotional equilibrium is brilliant in its simplicity. For each emotionally charged situation:

1. Recall your thoughts leading up to, during, and immediately after the situation.

2. Identify any distortions in your thinking through clear, logical analysis (skills that scientists use daily).

3. Replace the distorted thoughts with undistorted thoughts and positive self-talk.

These steps involve the skills of mental agility (awareness, letting go), flexibility (acceptance, analysis, creative problem solving), and core mental strength (affirming self). Again, the point of mental balance is the effective *integration* of these skills and application of this process; something that requires time, patience, practice, and gentle humor.

Behave "As If" You Feel Positive and in Control

A complementary strategy for maintaining emotional equilibrium is to behave in a way that is consistent with how we wish to feel, rather than how we actually feel, that is, to choose to act "as if" we are feeling centered and in control. Though it may seem counterintuitive, some have found it to be very effective. And it's quite amazing to observe (and experience) the results. (This approach also makes it easier for us to take our emotional reactions less seriously.)

Suppose, for example, that you are about to present an important seminar to your research group. You're not looking forward to it at all. You're uncomfortable speaking in crowds, and every time you think about it, you feel anxious. By viewing the presentation as an opportunity to talk about your exciting research with interested colleagues and choosing to speak with energy and enthusiasm, you will begin to *feel* energetic and enthusiastic. Your attitude and approach will engage the interest of your audience who will, in turn, begin asking probing questions that will help you to stay focused on your topic (rather than on yourself and how you feel) and stimulate you to further explanation.

"Ride the Wave" of Emotion

To reiterate: experiencing emotion is a normal part of being human. It is neither "good" nor "bad." It simply *is*. What is important is how we act in response. There will be times when strong feelings will persist well beyond our initial reaction, even when we have responded appropriately and effectively. For example, experiencing a loss—such as when you are not offered a job that you are perfect for and so hoped you would win—can engender a sense of grief that may take some time to subside. Though it may threaten to lock you in its grip (i.e., trigger reactivity), you are *not* helpless. You still have a choice. You can choose to "ride the wave" of emotion with acceptance, patience, and grace. This approach, like learning to take your emotions less seriously, is challenging but can succeed in helping you to maintain your mental balance through the difficult times. As much as possible, detach yourself from emotional reasoning. Choose to believe that there will be a time when you will feel like "yourself" again. Above all, do not interpret these periods as indicators that there is something "wrong" with you. Repeating phrases such as the following may inspire confidence and hope and help you to cope:

I'll ride the wave of this emotional storm.

I've coped well before; I can do it again.

This, too, shall pass.

In such situations, it is especially important to continue to affirm yourself through the support of trusted colleagues, friends, and/or family, positive self-talk, and other nurturing strategies.

Adopt a Healthy Attitude

Regardless of what other strategies you employ to maintain emotional equilibrium, above all, a healthy attitude is key. Accepting that feelings are normal and that everyone experiences them from time to time will reassure you that you are not alone in these experiences. Using positive affirmations, reminding yourself of your worth, and remembering your values and your commitment to succeeding in your chosen field will help you to shift your focus from any difficulties that you are experiencing to your goals and how to achieve them.

By viewing the situation from a larger perspective (e.g., of world crises/ life tragedies), you will gain perspective: by comparison, work problems will seem trivial. Focus on what is working (rather than not), what you have achieved (rather than on what is left to do), and your final goal (rather than a current setback).

Obviously, having a sense of humor about yourself and others contributes to a healthy attitude. Laughter is very therapeutic. And when used appropriately, humor can help everyone to take an emotionally charged situation—and themselves—less seriously. Humor can help us to put things into perspective and let go of negative thoughts and physical and emotional tension.

"Mental balance" is the ability to stay "centered" in one's thinking and feeling, to stay focused, and to think clearly—without distortion—so as to maintain sound judgment and emotional equilibrium. It involves integrating all other mental toughness skills and applying them effectively and appropriately in our daily lives.

PROPER NUTRITION

Just as providing appropriate nourishment to the body through healthful eating will support the physical development of an athlete, so too will "mental nutrition" enhance the development of every aspect of mental fitness. The techniques have already been introduced in other sections of this chapter and include adopting a healthy attitude (a "mental balance" strategy just discussed), living with integrity (making positive choices based on your values and beliefs), engaging in positive self-talk, repeating affirmations (mental strength), working for employers and in environments that are the best "match" for you, and choosing to associate with people who are supportive and encouraging (e.g., mentors and members of your network).

It is important to remember that our brains are part of our physical being. As well as supporting our *physical* development, eating healthfully, exercising regularly, and engaging in activities that are relaxing and restful (including having sufficient sleep) will also enhance *mental* function and positively affect our *emotions*. (An interesting and very practical book on the topic of the "mind-body connection," *Full Catastrophe Living* (Kabbat-Zinn, 2005), is based on the program of the stress reduction clinic at the University of Massachusetts Medical Center.)

MENTAL REST AND RELAXATION

Our physical bodies require regular periods of rest and relaxation (R&R) in order to recover sufficiently from training and prevent exhaustion and injury. So, too, do our minds. Not only do we need to take breaks from the (often intense) intellectual work of scientific research—for rest, social interaction, play—we also need to pause regularly from the intellectual and emotional work involved in developing mental toughness. No mental training program will be successful without it. If we do not allow ourselves sufficient time away from the rigors of mental work, we will suffer fatigue and (eventually) burnout.

You have learned how to stop negative self-talk, defuse your reactions, and moderate emotion during pinch periods and times of confrontation and conflict, but these strategies do not always completely defuse the associated tension. We need to develop other strategies for "letting off steam" and "recharging our batteries" so that we can approach each new day afresh, without the burden of unresolved emotion or mental exhaustion. For some circumstances, a good night's sleep is enough. In other situations, more active strategies are needed.

The nature and duration of periods of mental rest and relaxation will vary with your needs and circumstances. Do include in your "tool kit" strategies for short-term stress relief (that you can interject into your busy days), conflict/crisis relief (for dealing with mental tensions associated with stress), and long-term "mental health maintenance." Try different techniques to learn what works best for you in every situation. If you set realistic expectations and are patient with the process (and with yourself), you will succeed.

Take Short Breaks

Incorporating regular periods of R&R during your busy schedule is ideal. Scientific studies of exercise physiology have demonstrated the efficacy of short rest periods between sets of exercises in strength training. These "rests between sets" allow the working muscles to recover slightly, before the next set of exercises begins. This optimizes the training effect, while decreasing premature fatigue and risk of injury.

The same strategy can be applied to the working mind, with similar benefits. Our minds are capable of intense focus for a finite period of time (from 45 to 90 minutes, depending on the task, the individual and his or her training), after which we lose concentration. Forcing ourselves to stay focused after fatigue sets in will be counterproductive: the added mental and physical effort will result in a greater emotional deficit and the association of discomfort (and perhaps negative feelings) with the activity.

One strategy is to take short breaks from the intellectual work—even five minutes—to rest one's mind, change one's focus, and become more centered and mindful. Doing something as simple as straightening up from the computer screen, stepping back from the bench, closing one's eyes, and taking a few slow, deep breaths can work wonders. And it requires just moments of your time.

Occupational health and safety experts advocate doing gentle stretching exercises throughout the day to decrease the possibility of repetitive strain syndrome, back problems, etc. These stretching exercises, when combined with gentle, mindful breathing, and positive self-talk will also be advantageous. Other ideas include taking a brisk walk outdoors, chatting with a colleague, "surfing" the Internet, or playing a favorite computer game for a few minutes.

Focus on Something Else

Another strategy is to replace one activity with another. This requires a certain amount of mental energy, but the change will provide a break from the first activity and from the fatigue that could set in if pursued too long. More often than not, it will inspire renewed energy as you become motivated to push the second task forward. It will be encouraging to you to know that you are continuing to make progress, especially when you are feeling the pressure of looming deadlines.

The important point to remember is to shift your focus *before* you feel exhausted. Stop the first activity while you still feel positive about your progress and could continue a bit longer. Then, when you return, you'll feel more optimistic about beginning again.

This approach—engaging in a different, though equally demanding activity—is similar to "cross-training" in coaching and athletic training circles: one continues to advance one's level of fitness, while allowing the muscles, ligaments, and tendons emphasized in the former activity the time to rest and repair. Overuse injuries are greatly reduced.

Try this technique when you find yourself caught up in negative thinking or in reviewing negative conversations or situations. Gently, but firmly, set your mind on another topic or task. Though you may believe that nothing will help you to stop these negative thoughts, the simple act of *beginning* to engage in something else can be enough to break the obsessive focus. Picking up the latest scientific journal and beginning to read an article on your research topic, for example, will shift your attention to one of your areas of strength and will engage your curiosity and problem-solving and critical thinking faculties.

Engage in Physical Activity

Engaging in strenuous physical exercise is an excellent strategy for relieving the physical as well as the emotional tension associated with stress. The competitive aspects of an intense game of squash, for example, may provide the physical outlet that you need; your squash opponent becomes, metaphorically, your adversary and you can safely vent your energy.

If you prefer more solitary activities, running, swimming, and cycling are excellent options. Regardless of your choice of aerobic activity, you'll experience a pleasant sense of well-being afterward; the change in brain chemistry due to the release of endorphins gives one a natural "high."

Activities that combine gentle stretching, balancing poses, and mindful movement, such as yoga and tai chi, are also effective. And, as mentioned earlier, practicing these physical techniques can be an effective mechanism for developing other mental strategies (e.g., positive self-talk, letting go).

Talk it Out

Sometimes we simply need to talk about our experiences and feelings to someone who understands and cares. Speaking with a trusted mentor about our worries over career opportunities, for example, can help us to gain perspective and identify new options. Complaining to a sympathetic friend gives us a safe environment in which to release tension, frustration, and anger. Just knowing we are being *heard* can offer relief. And sharing the burden can often make it easier to bear.

Pursue Personal Interests

Most of us aspire to a life that combines professional *and* personal involvements. Though each of us must determine how we balance the two (see the chapter "Balancing Professional and Personal Life"), what is important is that we understand our values, goals, and needs and include activities in our schedules that will enable us to honor them. By doing so, the stresses in one aspect of our lives will be counterbalanced by the other parts of our lives that are satisfying. These activities are many and varied.

Engage in activities outside your professional life that have meaning for you and give you a sense of enjoyment. Read inspirational or escapist books if they take your mind off your work and make you feel good. Allow yourself unstructured time to pursue creative hobbies, such as music, art, and theatre. Engage in regular physical activity, practice meditation or other relaxation activities, listen to your favorite music, etc.

Many scientists who have children discover that attending to and focusing on the needs and activities of their young charges provide just the incentive that they need to take a break from work. As one female scientist put it: *"My children keep me sane. Without them, I surely would be a workaholic. As it is, having to take care of them provides a needed break and relief from the stresses of my working life."* For others, contributing to their community (social, religious, geographic) provides a tremendous source of satisfaction, affirmation, even comfort. Whatever your involvements, make sure they can be accommodated under your current circumstances; you don't want to put more pressure on yourself.

ENDURANCE

A final aspect of physical fitness is cardiovascular and muscular endurance: the ability to continue exercising at a given rate or intensity without fatigue. The benefits to physical health and athletic performance are obvious. Likewise, "mental endurance"—the patient, persistent application of all mental toughness skills—enables us to continue behaving responsively in the face of the inevitable intellectual (and emotional) challenges of our chosen profession. Well developed, it will give us lifelong resilience and enhanced professional and personal success and satisfaction.

Inspiration, motivation, and realistic goal setting contribute greatly to mental endurance and can be enhanced in myriad ways. These include developing a personal vision that is flexible (see the chapter "Career Management"), finding appropriate coaches and training partners and interacting with scientists outside your area of expertise (see the chapters "Mentoring" and "Networking"), reading outside your field (see the chapter "Continuing Professional Development"), and thinking globally and acting locally (see the chapter "Training and Working Abroad").

Discovering the techniques that work best for you is an ongoing process that requires time and persistence. But you will succeed. As Nadia Rosenthal concludes in the Prologue:

Patience is not the virtue I would espouse here, but rather a stubborn intolerance of personal compromise when it comes to pursuing your ideas. It takes clever strategizing to keep doing what you're interested in doing, in the face of shifting fashions and inconsistent funding. The politics and practicalities of research are necessary parts of the game, and can work just as well in your favor as against it. But the centrepiece has to be the science. If you are truly obsessed with a magnificent question, Nature never lets you forget it.

REFERENCES

Boice, R. *Advice for New Faculty Members*. Boston: Allyn & Bacon, 2000.

Burns, D. R. *Feeling Good. The New Mood Therapy (Revised and Updated)*. New York: Avon, 1999.

Kabat-Zinn, J. *Full Catastrophe Living* 15th anniversary ed. New York: Bantam Dell, 2005.

Murray, W. H. *The Scottish Himalayan Expedition*. London: Dent, 1951.

M. Elizabeth Cannon, Ph.D.
Professor and Head, Geomatics Engineering
Schulich School of Engineering, University of Calgary, CANADA
Geomatics Engineer

Dr. Elizabeth Cannon is an internationally recognized researcher in the area of satellite navigation systems. She is the first Canadian woman to receive a Ph.D. in geomatics engineering and the first female faculty member in the University of Calgary's geomatics engineering program. Dr. Cannon has conducted research in the field of the Global Positioning System (GPS) since 1984, in areas such as vehicular navigation, precision farming, and aircraft flight inspection. Her commitment to the application of her research through technology commercialization has led to the licensing of software to over 200 agencies worldwide. Her leadership skills have led to numerous positions within various associations, including President of the U.S. Institute of Navigation (USION)—the second woman and first non-American to hold this position.

The impact of Dr. Cannon's work has resulted in her being named one of Canada's "Top 40 Under 40." She was elected as a Fellow of the Canadian Academy of Engineering and the Royal Society of Canada and is the only woman to receive the Johannes Kepler Award (in 2001 from the USION)—for her outstanding contributions to satellite-based navigation.

Dr. Cannon has long been dedicated to the encouragement of women in science and engineering. As the 1997–2002 Natural Sciences and Engineering Research Council (NSERC)/Petro-Canada Chair for Women in Science and Engineering (Prairie Region), she focused on raising public awareness about opportunities for science and engineering careers and conducted research into the reasons why men and women choose these careers and the factors influencing their decisions. She also developed innovative programs to support the women themselves, such as an e-mail mentoring program (SCIberMENTOR), which she continues to lead.

Her partner at home and work is Dr. Gérard Lachapelle, whom she met through working in the satellite navigation field. She considers herself fortunate to be able to share with him both the excitement and discovery of academic life and their two children, Sara (age 17) and René (age 14)—who make certain she doesn't take her work too seriously.

KATHLEEN SENDALL, MBA
Senior Vice President, North American Gas
Petro-Canada, CANADA
Mechanical Engineer

Kathleen Sendall is an outstanding leader in the North American Oil and Gas industry. A graduate of mechanical engineering (B.Sc., Honours, Queen's University, Canada), and the Ivey Executive Program of the Richard Ivey School of Business (University of Western Ontario, Canada), she joined Petro-Canada early in her career as a process engineer working on a proposed LNG project. After a two-year hiatus with Nova Gas Transmission, she returned to Petro-Canada, and in 2000, became the organization's second female vice president. She is responsible for the company's North American Natural Gas Business Unit and oversees the work of approximately 1,000 employees.

Ms. Sendall's service to engineering and society is exemplary. She is President of the Canadian Academy of Engineering, a director (Executive Committee) of the Canadian Centre for Energy Information, a director of the Canadian Academies of Sciences, and a director of Calgary Opera. She sits on the Board of Governors of the University of Calgary and is President of the Calgary Chapter of the International Women's Forum. Ms. Sendall is a governor of the Canadian Association of Petroleum Producers (CAPP) and currently chairs their CEO Task Force on Climate Change. In April 2005, she became the first woman to be Vice Chair of CAPP.

Her significant contributions been recognized by many awards and honors. She received the YWCA Woman of Distinction Award (1998), was named a Fellow of the Canadian Academy of Engineering (1999), was one of *"The Globe and Mail's* Top 100" (2003 and 2004), was also one of, "the [*National*] *Post's* Power 50" (Canada's top businesswomen) (2002), and was awarded an honorary Bachelor of Applied Industrial Ecology degree from Mount Royal College (2002). In 2004, she received a Summit Award for Community Service from the Association of Engineers, Geologists, and Geophysicists of Alberta. To these accomplishments, Ms. Sendall adds another, more personal one: "I am very proud of the fact that I have been able to successfully balance a career and family and have raised two wonderful children."

8

· · · · · · · · · ·

PERSONAL STYLE

· · · · · · · · · · · · · · · · ·

M. Elizabeth Cannon and Kathleen Sendall

> *Have you ever noticed a woman walk into a room and wondered who she is, knowing she must be someone of significance, just because she looks impressive? Someone who exudes confidence and professionalism, even before she speaks? Without saying a word, that woman is communicating to you.*

Whether we are being intentional about it or not, all of us send messages to everyone we interact with, *all* the time; verbal and nonverbal "signals" that influence how others perceive us, our ideas, and our abilities. For working women scientists, it is essential to understand these messages and ensure that they are the ones we intend to send—ones that reinforce our competence and capability—without comprising who we are as individuals. We need to understand "personal style."

Editor's Note: I wish to thank Dr. Karen Spärck-Jones, Professor of Computers and Information (Emerita), University of Cambridge, for first introducing me to the terms "surface" and "functional" style, which capture so perfectly my original concept for the following three chapters and represent important threads that weave through the entire book. Her unwavering commitment to excellence and intellectual rigor—as a scientist, educator, and mentor—have been a constant inspiration.

What is Personal Style and Why Does it Matter?

It takes anywhere from six to 60 seconds for someone to form an impression of you; the most commonly cited time is 30 seconds (Bixler and Nix-Rice, 1997). This is hardly longer than it takes to say "How do you do?" Yet a critical first impression is formed. And once formed, it is difficult to override. As humans, we tend to group perceived similar characteristics together (McCoy, 1996). So, when we see a woman such as the one described above and perceive her to be professional, confident, and well groomed, we also are likely to consider her to be competent, successful, and organized and to have great interpersonal skills. And if our initial perceptions are highly positive, we are likely to retain them for a period of time, even when subsequently confronted with evidence to the contrary.

Conversely, when we meet someone for the first time who is wearing sloppy or inappropriate clothing, with messy hair and slouching posture, our critical, first impression will be dominated by our *perception* that they may also be disorganized and incompetent and may fail to pay attention to detail. They will have to work harder to create a *positive* impression (than the woman we just discussed) because our initial perception—which was negative—will be difficult to overcome.

Many of us rebel against the notion that "appearances count"—as scientists, we are conditioned to believe that "results count"—but clearly, *both* count. The old saying, "Don't judge a book by its cover" is undoubtedly wise, but in today's world, we cannot deny that physical presence is vitally important in influencing respect. Failure to recognize this crucial fact can sabotage our professional relationships and opportunities for advancement and even undermine the acceptance and credibility of our work.

Knowing that we are projecting a positive image also has a beneficial effect on our performance. It is no coincidence that advertisers focus on "looking your best and feeling your best." When we know we are looking our best, we tend to feel more confident, and when we feel more confident, we are more likely to take on new challenges and excel at them. Our personal style affects not only how others see us but also how we view ourselves.

We don't mean to suggest that all professional women need to appear cloned—it is critical to develop your own personal style and adapt that style as necessary, given your role and your working environment. Your personal style should never mean "uniformed." Different work environments, locale, cultural norms, as well as the task being performed, all need to be considered.

THE COMPONENTS OF PERSONAL STYLE

Be aware of your "default" personal style and its impact. Notice your style under pressure. Ask respected colleagues and trusted friends what they have noticed about you and your interaction style. If you are not sure what it is, have a personal style inventory done by a professional.
                                    Karin Porat, Executive Coach, Calgary, Canada

There are two distinct elements to personal style. The first is *surface style*, which consists of *visual* elements (primarily dress and posture), and second, *functional style*, which is a combination of *nonverbal* and *verbal* elements (facial expression, body language, tone of voice, vocabulary). What is interesting to note is that these two aspects of style are far from evenly weighted in the formation of that critical first impression. The visual elements account for the majority of the impact you have on others at first meeting.

Surface Style: What You Wear and How You Look Count

Clothes make the man; naked people have little or no influence on society.
                                    Mark Twain

Clothes also make the *woman*. Merely covering your nakedness is not enough to command respect as a professional.

The most immediately visible aspect to your *surface style* is dress. How you dress is your "packaging"—compare the amount of time, effort, and resources that product manufacturers spend on sending the right message to the consumer through their packaging, to the haste with which many of us shop and dress. Ask yourself: "Are consumers more likely to be attracted to cereal in a great package compared to one in an ugly brown box, especially if the nutritional composition is the same? Why should I think about my 'personal packaging' any differently?"

What Are the Options?

In virtually every workplace today there are three accepted ranges of style, with some variations, depending on one's location and culture.

Formal work attire. Whether you are a research scientist or a bank executive, this is an important style to understand and be comfortable in. Whereas a corporate executive's environment may require this style of dressing every day, the research scientist may reserve this style for important internal and external meetings, presentations, and job interviews. This is "high-end" dressing. Every woman needs at least one high-quality suit in black, navy, or gray; a crisp blouse; tasteful jewelry; and supportive, medium-heeled shoes. It is not necessary to appear like a "surrogate male"—by all means, develop your own sense of style—but it is important that the effect be understated, not flashy.

General work attire. This is the equivalent of our male colleagues' sports jacket, slacks. and tie and is appropriate for most day-to-day activities in the work environment, such as less formal, internal meetings and teaching. The look is somewhat more casual, with pants and jackets in a variety of colors being appropriate.

If most of your work is in the laboratory or in the field, the definition of "general work attire" needs to reflect a style suitable for these environments. For laboratory work, comfortable clothing with low heels would be appropriate. A jacket would not typically be needed if a lab coat is worn, as too many layers may be restrictive. A nice blouse or sweater would be a good alternative. For the field, layers of clothing are recommended (depending on the weather conditions) and could include a high-neck sweater, light wool or fleece jacket, and a windbreaker or overcoat. Khaki (or dark) pants and footwear that is safe and durable would finish the look. This style is further discussed in the section on matching an appropriate personal style to the position, function, and cultural context.

Either way you define "general work attire," the look must appear "finished," with careful grooming. And save the trendy outfits for after hours when you are with family or friends.

Casual work attire. Perhaps nothing has generated as much confusion in the workplace in recent years as the growing trend toward more "casual" dressing. As noted above, the choices for women are particularly bewildering, and a quick scan across a university campus will provide ample evidence that many women may be sabotaging their professional image by the alternatives that they are choosing. The extreme casual end of the workplace attire spectrum is dangerous territory—you run the risk of being perceived as having an entirely different role than the one you have. It is a fact that women cannot dress as casually as their male counterparts and maintain the same level of professional image (see story below).

Standards Are Different for Women

Although our male counterparts lose some of their perceived authority when they dress more casually, the impact on their female counterparts is much more significant. In his book, *New Women's Dress for Success*, author John Molloy (1996) provides a startling example:

We showed pictures of a man and a woman of normal height and weight with the same coloring wearing similar suits. The woman wore a skirted suit and the man wore a traditional menswear suit. We asked a cross section of business people to guess how successful they were and how good they were at their jobs. In most surveys, the man had a slight edge. But the edge was very small, and it is shrinking every year. When you have both parties remove their jackets, the man wins hands down. When we show his picture to audiences, 80–90% always assume that he has a jacket somewhere. They also believe he graduated from college, was good at his job, and was an executive or professional. When we show the woman without a jacket, close to 80% of business people assume that she did not take off her jacket. They also assume that she is not an executive or professional, but a clerk, a typist, or a secretary.

Is this fair? No. Is it a reflection of reality? Yes. Women, particularly those working in male-dominated fields such as science, need to dress more formally than their male colleagues. Here are some specifics.

Rarely wear jeans. Even if the formal or informal dress code in your workplace includes denim, jeans are rarely a good idea. If you are in the small percentage of the population who are fit and toned *and* look great in jeans, you may look too sexy. Men enjoy *looking* at women who dress sexily—however, they don't *promote* them—and other women are often disapproving. For the remaining 99% of us with imperfect bodies, jeans rarely look attractive or professional (Bixler and Nix-Rice, 1997).

Wear a jacket. A jacket instantly creates a more formal look and commands more respect. A good tip is to always have a blazer hanging in your office for those unexpected encounters when you need to look more polished.

Maintain good grooming. Grooming is never more important than when dressing casually. A good haircut or hair pulled off the face, tasteful makeup, impeccable hygiene, and excellent footwear instantly create the image of a woman who commands respect.

Dress casually wisely. In a research or teaching environment, casual wear is acceptable. But this type of dress may not be suitable some of the time. Many of our female colleagues have found that for a woman, wearing a jacket sets a professional tone that may be required to show that she should be treated with respect. This may be particularly important in cases when she is meeting an audience for the first time and her reputation does not precede her. An obvious example is the female faculty member who has to teach a large, first-year class: she will be somewhat unknown to the students because they are new to the department. And since teaching large classes includes controlling the class, it is critical to create a positive impression.

> *I learned firsthand the impact of physical presence on classroom management when I taught my first lecture of an introductory engineering class. When I walked into the room of 160 students, four of the men jumped out of their seats, convinced that they were in the wrong class. Clearly I had not established a high degree of credibility at first glance!*
> Elizabeth Cannon, Professor, University of Calgary

For good or ill, students *do* watch what you wear.

> *Every year I receive comments about what I wear on my teaching assessments, usually about how I wear too much black; in contrast, my male colleagues have never received comments on their clothing choices. I find it remarkable that my black sweater and slacks create more cause for comment than my male colleague's kilt! Rightly or wrongly, the dress and style of a female faculty member make an important impression and will be noticed.*
> Jocelyh Grozic, Assistant Professor, Faculty of Engineering, University of Calgary

Senior level students, on the other hand, often know of a faculty member's professional accomplishments so she can garner credibility from her work. She can then walk into the classroom having already created a positive impression.

Taking care to dress wisely may be even more important when you start your career since you may be significantly younger than most faculty members; many new women professors have been assumed to be students. Credibility has to be earned, and a polished personal style will at least set the stage for a good first impression.

Dress carefully for after-hours events. Most women scientists are invited to attend events after work hours. Some of these are directly work—related and may include a reception on the occasion of an important scientist visiting your institution or company, or a dinner to celebrate student awards. In these cases, the required dress is formal work attire or, perhaps, general work attire (but not your field clothes!) if the event is not expected to be too formal. Check with the host if you are not sure of the dress code.

Other events are those that appear purely social—the office Christmas party and a colleague's birthday party are two examples. In these cases, the invitation may explicitly state the dress code so you can gauge what to wear. However, since your colleagues will be in attendance, you must be careful about what you wear; these events really are extensions of your workplace and you will be judged accordingly. Do not wear anything too revealing that you may regret the next day. If the event is black tie, a classic long dress or formal pantsuit would be good choices. A formal ball gown with frills and a full skirt would generally not feel comfortable at a social event with professional colleagues.

Physical "Presence" Is Important

There are two other important contributors to your surface style: posture and weight. Very few work environments are populated by women (or men) with perfect, slim bodies. Yet it is a fact (however unfair) that, in North American society at least, slenderness is equated with success, affluence, organizational ability, and discipline (McCoy, 1996). Unfortunately, there is a perception that overweight women are less professional; so those of us who carry extra weight have to work more diligently to cultivate a professional image. Well-fitting, monochromatic clothes and careful grooming are necessary elements, as is a natural, upright posture. "Good posture not only takes off 10 years and 10 pounds, it creates an instant impression of competence" (Bixler and Scherrer Dugan, 2001). Even the best attire won't look professional on a slouching body. Of course, achieving and maintaining a healthy weight and good posture through diet, exercise, and practice are important for many reasons other than personal style (e.g., it enhances our ability to manage anxiety and stress and contributes to mental toughness and resilience)—we just know that it requires dedication, patience, and persistence.

Summing Up Dressing Up

It is often suggested that you take your cue in how to dress appropriately from observing how others dress. This may be a difficult challenge for those of us who work in male-dominated environments where there is a dearth of female role models. One strategy is to observe other successful women in similar environments and use their approaches to dress as a clue. Another is to consult the Internet and print media for some guidelines. (Some useful references are provided at the end of the chapter.) Below is a set of useful guidelines based on those offered by Molloy (1996):

1. Be one of the more professional and conservative dressers in your group.

2. If your boss is a man, be as traditional as possible. If your boss is a woman, mirror her level of formality (without copying her outfits), while maintaining your own style.

3. Stick to colors and color combinations found in traditional men's sportswear; add a dash of color to your outfit if you want to add some individuality.

4. Pay attention to your accessories: briefcase, pen, etc.

5. Make sure that your hairstyle and makeup say "professional."

6. Wear nothing that might be considered "too sexy for the office" by your parents. If you work for or with men, have a man whom you trust check your casual work attire. If he tells you an outfit that you think is conservative is sexy or inappropriate, listen to him. His perspective may be shared by others.

7. Choose natural or natural-looking fabrics that do not wrinkle easily. Wrinkled attire will make you look disheveled.

8. Wear "serious" footwear—no running shoes, sandals, open-toed party pumps.

9. Keep a navy or black jacket handy that you can wear to unexpected meetings.

10. If you wear pants, make sure that they are well tailored and full cut.

11. Don't try to copy the outfits your male coworkers wear.

12. Neatness counts more when you are dressed casually. Check your clothing, hair, and makeup regularly.

13. Posture is important. If your posture isn't perfect, practice and exercise in order to improve it. Well-cut suits can camouflage imperfect posture; most casual outfits draw attention to it.

Of course, as with any set of guidelines, there will always be exceptions. Some tremendously successful women have developed a personal style that does not follow the general rules.

I'm often told that I am the most casual dresser among academic administrators that anyone knows. Jeans with a T-shirt or fleece sweater is my preferred style and I dress this way as often as possible. Stories abound of my wearing jeans at occasions where almost everyone else was in suits. The most embarrassing was the time I had to borrow a jacket from a faculty candidate for a formal photo shoot that I had forgotten was on my schedule. Sigh...When I add to this that I paint watercolors whenever I am in day-long meetings of boards or committees, I sound like someone who just takes pleasure in being different from others. While that's undoubtedly part of why I behave this way, this style works well for me for other reasons.
Maria Klawe, Dean of Engineering, Princeton University

These guidelines are just that: suggestions to use at your discretion. Never compromise who you are for the sake of following someone else's advice. If you want to reflect a style that would not be considered mainstream, it is advisable that you do so with care and attention. What works for one person may not work for someone else. Generally, as you become more established and have earned a high degree of credibility for your work, you are more able to get away with bending the rules.

Keep in mind that your *surface style* has the biggest impact on the critical first impression that others have of you. So make sure that you give it due diligence. Although a successful career for the woman scientist does not depend entirely (or even primarily) on developing an appropriately professional personal style, research suggests that *not* doing so can be a primary contributor to *failure*.

Functional Style: What You Do and What You Say Count

Once that critical first impression is formed by your *surface style*, it is your *functional style* that subsequently reinforces (or not) an initial positive or negative perception. As noted earlier, it takes much more effort to overcome a negative first impression.

Functional style is a combination of *nonverbal* and *verbal* elements (facial expression, body language, tone of voice, vocabulary). It is often alluded to by descriptors such as "presence," "demeanor," "personality," or even "charisma." Bixler and Scherrer Dugan (2001) point out that:

> DR. ALBERT MEHRABIAN, in his famous body language studies at UCLA, found that only 7% of the emotional meaning in a message is composed of the actual words we use. About 38% is communicated through the tone of our voice and voice inflection. About 55% comes through our non-verbal communication, which includes facial expression, gestures and posture. This startling statistic reminds us that others believe the visual information that we make available to them before they believe the actual content of the words we use. Even more profound is that all of us believe what we think we see, before we believe what the communicator intended to project.

Nonverbal Aspects of Functional Style

Eye contact. In North America, eye contact is critical. We instinctively distrust those who fail to make direct eye contact with us (Bixler and Scherrer Dugan, 2001). The eyes are considered the "mirror of the soul" and direct, steady eye contact indicates interest and promotes trust. Maintaining eye contact in a discussion is the strongest way to express engagement in the conversation. We all have interacted with individuals whose eyes are constantly shifting or, worse, who are constantly looking past us (perhaps for someone whom they consider worthier of their attention?). In both circumstances, the individual creates a strong, negative impression. Interestingly, we have observed that women have no difficulty in establishing and maintaining eye contact with other women but often have to make a conscious effort to do so with their male colleagues.

It should be noted that the cultural context is important. In some Asian and Middle Eastern countries, for example, making direct eye contact is considered disrespectful. It is important, therefore, to do your research each time you travel abroad or interact with people from other countries. You need to understand the cultural context of your interactions (Bixler and Scherrer Dugan, 2001).

Body language is another powerful signal of engagement. Leaning forward, with a centered, open body posture and a pleasant facial impression, indicates that you are receptive, open, and actively listening. A slouching, off-center posture, combined with crossed arms and an unsmiling face, suggests a lack of interest, boredom, distrust, and negativity (Bixler and Nix-Rice, 1997). Head nodding when listening to someone speak is often used by women to signal that they are listening. But it can be misinterpreted as *agreement* with what is being said. Tilting one's head slightly to one side also indicates attention. But beware: excessive nodding or head tilting can give the impression of submissiveness; this can erode your credibility.

> *Many careers have been sabotaged by nonverbal "ticks" and habits. The fastest way to change the impression you are making is to adjust your nonverbals: dress, voice, posture, and facial expression are key to perceived competence and confidence. Monitor and adjust to fit the situations.*
> Karin Porat, Executive Coach, Calgary, Canada

Use of space is not often thought of as an element of style, but it is important and has important gender and cultural considerations. In North American culture, individual space is generally thought to have a radius of three to five feet (Bixler and Nix-Rice, 1997). Because of their physical stature, men automatically occupy more personal space than women, and they use it differently. Observe most men sitting in a chair around a conference table: they tend to stretch out their legs, drape an arm across the back of an adjacent chair, or clasp their hands behind their heads with their elbows out. In contrast, their female colleagues often sit with their legs and arms crossed. Men's use of physical space conveys an image of personal power and relaxed confidence.

In North America, personal space tends to be breached by only the handshake. Women who constantly touch others (particularly men) in the course of conversation (often quite unconsciously) may be perceived as being aggressive or, worse, *sexually* aggressive.

The norms regarding personal space vary considerably from culture to culture. In one culture, enthusiastic kissing on each cheek is considered acceptable, while no touching whatsoever may be the norm elsewhere. Again, it is vitally important to do your research.

Gestures, such as hand and arm movements, also send important nonverbal cues. Used appropriately, gestures can effectively punctuate the spoken word. It is critically important that we do not sabotage our message of confidence with nervous, anxious gestures such as twisting a ring or playing with our hair. Large, calm, deliberate gestures tend to convey confidence. Since gestures tend to be subconscious, it is wise to concentrate on trying to be as still as possible, using gestures only minimally until you are well practiced at using the gesture effectively.

Given the importance of nonverbal communication in defining our own functional style, it is worthy of time, attention, and practice. Try having someone you know videotape you, both in conversation and making a presentation or teaching. Identify jointly those behaviors and habits that could be creating a negative impression and focus on correction and improvement. And identify those that create a strong, positive impression and use them more deliberately.

Verbal Communication and Oral Presentations: Say It "Right"

Verbal communication is a vital aspect of what scientists do (as you'll read in the chapter dedicated to this topic). Most jobs entail participating in meetings and conferences, giving presentations (technical or strategic), teaching, and/or training. It is imperative, therefore, that your style be clear and effective and project confidence. This is particularly true for women in male-dominated environments where gender differences in communication style may increase the possibility of misunderstandings in an already complex process.

Functional style is important in meetings within our organizations, professional associations, and collaborations. In addition to ensuring that we use body language to our advantage (e.g., by the way we sit and use gestures), it is important to project confidence and competence by participating in discussions and offering advice on solutions to problems. A fine balance must be struck between contributing to the group effort and talking too much; if your colleagues perceive you to be monopolizing the conversation, for example, or to be continually advancing the same agenda, they may become irritated and stop listening. And if this reaction becomes habitual, it will be very difficult for you to have your ideas "heard." One senior scientist, whose enthusiasm for contributing in meetings became a concern for her, developed an interesting and effective way of holding back. Though it is not an approach that we'd recommend to early career scientists—you need a well-established reputation and a solid track record of success for it to be accepted—it is a perfect solution for her:

I began painting in meetings several years ago because weekends are traditionally my painting time and I was spending too many weekends in meetings. I soon realized that painting has a wonderful side effect. It keeps me quiet. I have always had a tendency to talk much too much at meetings. Painting allows me to follow everything in the meeting but is absorbing enough that I do much better at controlling my need to talk. And it makes every meeting enjoyable.

Maria Klawe, Dean of Engineering, Princeton University

In their many years of participating in scientific conferences, the authors have observed women giving technical presentations on their work. In most cases, they were excellent. In some, however, the presenters' styles detracted from their messages. Some of the common mistakes included (1) speaking too softly and in a monotone voice (a microphone can only do so much), (2) using language that gave the impression that the speaker was not confident in her knowledge (e.g., "I guess you are right..." in answer to a comment from the audience), (3) standing in one spot rather than moving on the stage (when using a wireless microphone), and (4) not making eye contact with the audience.

These can be redressed in a number of ways. All involve honing your skills through education (there are many seminars and workshops being offered) and practice (in front of "friendly" audiences; you may even wish to have yourself videotaped), and learning from the feedback of others (some organizations and institutions have instructional development centers). (The chapter "Communicating Science" discusses the topic in greater detail.)

Scientists are trained to be critical thinkers. This means that we become accustomed to passing judgment on someone else's work. Some find this difficult; others seem to enjoy it! Regardless, it is essential that you develop skilful and effective ways of defending your work with clarity and confidence and not take the criticisms personally (see the chapter "Mental Toughness" for techniques on the latter).

Nowhere are the gender differences in personal style more evident than in *verbal* communication—in fact, entire books have been written on the subject. John Gray's *How to Get What You Want at Work* (2003) is one of the best, and his "101 Ways for Women to Score Points with Men" (p. 274) is a useful reference for those of us in predominantly male workplaces. It is important to remember a few of key points that contribute to developing a powerful, professional communication style:

Speak directly and factually. Be solution-oriented in your content.

Learn to project your voice. Unless your audience can *hear* what you are saying, you won't be able to make your point.

Use appropriate vocabulary. People with broader vocabularies are perceived to be more professional.

Practice speaking in a well-modulated tone. Use changes in volume to emphasize points. Control the pitch of your voice—the higher the pitch, the more "strained" you may sound (even if you're not). The lower the pitch, the more confident and controlled you will sound. Avoid habits like raising your voice tonally at the end of a sentence.

Learn the power of silence and the "pregnant pause." Don't rush to fill quiet moments in the discussion.

Avoid "ums" and "uhs" and other "filler phrases." You may wish to use videotaping to assess your speaking style or seek help from a coach, if necessary.

Understand how your listeners use the language. Although we may choose the same words, men and women quite often use and understand them in very different ways. (This problem also occurs in cross-cultural communication.) Learn to recognize when this is happening and make adjustments accordingly.

Be enthusiastic. Be energetic. Be positive.

Though we can inadvertently create a negative impression by having a personal style that projects a lack of confidence, we need to be equally cautious of appearing *over*confident. Men generally tend to be able to get away with being brash and overly passionate about their work, but we are not. In the authors' own experiences, women (including ourselves) have to walk a narrower line when it comes to acceptable behaviors. We have seen cases in which "Jim" flies off the handle when discussing a particular technical detail, and the outburst simply is brushed off with a comment: "Oh, that's just Jim!" But a woman behaving the same way does not receive similar treatment and may even be perceived as aggressive, erratic, or emotional.

Functional style is equally important in academic environments, especially in teaching. Studies have revealed that for women, teaching in a male-dominated faculty such as engineering poses many more and different challenges than for men. Krupnick (1985) demonstrated that, to be successful lecturers in an engineering faculty, women not only had to be technically competent in their subject matter, they also had to exude the stereotypically ``female" characteristics of being nurturing and approachable. That is, they had to be perceived as caring for the students and their progress (nurturing) and exhibit behaviors such as smiling in class and when greeting students in the hallways (approachable).

> *I have learned from personal experience that women may have to conduct themselves differently in the classroom. Some of my male counterparts could appear aloof or even tough ("if you don't settle down, I will throw some of you out of class"), while I could not. We are a curiosity as it is, let alone if we display characteristics that are not very familiar to*

> *young men (even if it is the 21st century). My goal was to develop a functional style that was not artificial but that was effective.*
> Elizabeth Cannon, Professor, University of Calgary

In summary, although *surface style* creates the critical first impression, it is your *functional style* that will firmly establish the impressions others have of you in the long term. Understanding each of the elements and identifying areas for improvement is time well spent in a professional's career. And by continuing to evaluate and learn from our experiences of interacting with others, we will become more confident and effective in our ability to work productively with others.

> *Notice things about the people you interact with: how they process, make decisions, react to pressure, their "hot buttons" and biases, what works and doesn't work when interacting with them, etc. Record what you notice. Update your files regularly. Refer to them when planning for important interactions.*
> Karin Porat, Executive Coach, Calgary, Alberta

MATCHING PERSONAL STYLE TO ONE'S POSITION, FUNCTION, AND CULTURAL CONTEXT

Personal style will change and develop with your career (and age) and may be particularly driven by the positions that you hold or the various functions that your job entails. For example, in many cases women scientists may have to do field work that would typically mandate attire that is appropriate, i.e., comfortable, safe clothing that may include a hard hat or steel-toed boots. Laboratory work may also require a lab coat and safety glasses. These may not win a style award, but they are appropriate and may be matched with clothing that can withstand potential spills or stains. Teaching may require low-heeled shoes for comfort when standing for long periods of time, accompanied by a jacket, to maintain a sense of professionalism and authority with the students. As previously mentioned, wearing a jacket may be particularly important in gaining credibility in a large, entry-level class where you may be a relative unknown to the students.

There is a saying that *"You should dress for the job you want—not the one that you have."* This generally applies well to the corporate sector, where there tends to be a clear hierarchy (e.g., entry-level engineer, senior engineer, supervisor, manager, vice president, CEO). In this environment, progressing in one's career involves moving up in the hierarchy. Your personal style will change during your career transitions, although the basic components discussed in the previous sections will be threaded throughout this evolution. In many cases, the company may have a dress code that will help guide your choices.

In the past five years, we've seen some divergent approaches to dress codes and work environments within the corporate sector. For example, some information technology (IT) companies almost *promote* a relaxed, carefree dress environment (some offices are even equipped with game tables), while many of the larger multinationals are sticking to a more formal work environment (with, perhaps, a "casual Friday").

In the academic sector, the hierarchy is somewhat blurred, as you'll read in the chapter "Climbing the Ladder." There are different "tracks" that one can take: one that emphasizes research and teaching roles and another that entails administrative duties, such as would be performed by a dean, provost, or president.

On the research and teaching track, the job functions entail developing and delivering course work, leading a research team, and providing service to your institution, profession, and community. In this capacity, you also interact with students (in the classroom, your office, and the hallways) and other faculty (on committees or on collaborative projects). Since these functions more or less stay the same over one's career, a personal style developed early in your career can be appropriate and effective for the long term. Academic institutions are very tolerant of a range of surface styles, from jeans with sandals to suits. The key is to develop one that works for you and allows you to get the job done. By using some of the guidelines in the previous sections, you can develop an appropriate style that you call your own.

There will be special cases when more formal dress would be appropriate, such as when attending a conference. In these situations, it is recommend that you dress neatly and professionally since you will be meeting and interacting with people you may only see infrequently. Any impression that you create will be remembered for a long time.

Another function that has typically been added to the academic's job profile is meeting with potential research sponsors. This is more of a *marketing role*, so you'll be most effective if you present yourself as someone in whom a company or agency would want to invest. The authors often have representatives of companies and government agencies visit their facilities to discuss collaborative research and/or to tour the laboratories. In these cases, professional dress is a *must*. Even research associates are expected to be professionally dressed (i.e., neither jeans nor T-shirts).

If your career choices lead you toward administrative duties, your dress will be more formal, with a nice jacket with pants or a skirt. Academic administrators are increasingly being expected to liaise with industry, research agencies, and government (including political) leaders, all of whom dress formally. Even for those who prefer more casual dress, it is important to know when to adjust your dress to suit the occasion.

In the first couple of decades of my career, women who dressed in suits or dresses in my research fields (computer science and mathematics) were mistaken for secretaries or representatives of book publishers. Women researchers dressed in jeans. When I became an administrator, I continued to dress casually, as a way of stating that the culture of being a serious researcher was more important to me than that of being an administrator. This seemed especially important to me since academics who become administrators are often assumed to instantly lose all technical interest and ability. (Of course, I have to wear dressier clothes occasionally.) Over time I have found that colorful patterned jackets and shirts over silky black stretch pants are comfortable, relatively resistant to the wear and tear of my working life, and dressy enough for most academic occasions.

MARIA KLAWE, Dean of Engineering, Princeton University

Interestingly, once you have developed a style that you are comfortable with and that you project regularly, it becomes part of defining who you are.

My style is typically suits (skirts or pants) with high heels, so that's what most people expect me to wear. The odd time, I wear informal clothing (pants with loafers and a nice shirt) when I do not have any meetings or I am trying to finish a project or paper. In these cases, I notice that I can walk by a colleague and they hardly notice me since my dress and my person seem to be mismatched. Since my goal is to get a job done on a deadline, I don't mind fading into the woodwork.

ELIZABETH CANNON, Professor, University of Calgary

Understand the Cultural Context

When it comes to developing our personal style (surface and functional), we are significantly affected by the norms and expectations of the society in which we live. Despite the fact that globalization has made us aware of other cultures, it is very important that we fully understand the values, practices, and verbal/nonverbal signals that exist in other parts of the world. This is particularly true in science, where technologies and ideas are exported and imported and international relationships are typical. This is amplified in academic environments due to the existence of international conferences, international societies, and collaborative research projects that may span several countries and cultures. Academic institutions also attract students from many countries (in some cases, over half of the graduate students in a research group may be from other countries). These students need to adjust to the cultural norms of the host institution, just as each member of the research group needs to be able to function in a multi-cultural research setting.

The potential impact of cultural differences on effective communication and the establishment of good working relationships can be significant. When these differences are overlaid with the fact that you are a woman scientist, your interactions can become that much more complex. It is prudent, therefore, to do some research when preparing for international travel or hosting colleagues from other countries. You may find it helpful to speak with others who have traveled to the country in question (particularly women). Their experiences, insights, and suggestions can be most instructive.

An excellent reference book you may wish to refer to is *Kiss, Bow, or Shake Hands. How to Do Business in Sixty Countries* (Morrison et al., 1994). For each country, the authors provide background information (history, political situation, language, religion, demographics) and cultural overviews (description of cognitive styles, negotiating strategies, and value systems), as well as describe expectations for business practices (appointments, negotiating, entertaining) and protocol (greetings [handshakes versus polite hellos], forms of address, gestures, gifts, and dress).

In Japan, for example, the style tends to be quite formal. Engineers working for companies wear suits at work, despite having to hook up equipment and climb ladders. At academic institutions, the dress is less formal, although the senior professor generally dresses in formal or general work attire. The forms of address also tend to be quite formal and it can take a while before they will call you by your first name, if at all (Morrison et al., 1994). This is particularly true if you are a woman. Unlike in North America, where coffee breaks are typically on the run, the Japanese scientists will often take a formal break and have tea served by a support staff member.

Gifts are not uncommon and may include some items with the company logo on it, or they may be more elaborate (the authors have received scarves and perfume). It always is good practice to take some gifts with you to reciprocate, although they need not be expensive; books showcasing your local environment or items from your university bookstore are suitable.

There are very few women in science careers in Japan (in companies and academic institutions), so most of the women whom you will encounter will be support staff. Regardless, professional women visiting Japan to conduct joint work will be treated with great respect. It is common for Japanese men to spend some time together after work at a restaurant or club, which makes their days very long (the average work week alone, excluding after hours get-togethers, is 48 hours [Morrison et al., 1994]). In some cases, it may not be appropriate for women to join men in these outings, so try not to feel left out or neglected if you are not invited.

Since there are so many countries and cultures that make up Europe, there is no one style that defines the region. Conventions and expectations are somewhat similar to those in North America, though generally more formal (in both surface and functional styles). Overall, there are fewer women in science leadership positions (in academic institutions and corporations) than in North America, so some men are not accustomed to dealing with female colleagues. In some cases, this can translate into awkwardness in interactions. For example, it is not considered prudent for a man to pass comment on a woman's dress in North America (even if it is a compliment), whereas in Europe, this may be viewed as acceptable, perhaps even expected. Generally, the concern about political correctness that has permeated North America is just starting to catch on in Europe. This can cause some uneasiness for women scientists traveling to Europe.

> I learned about the differences that can exist in other countries when I was being introduced before I was to give a lecture to a group of graduate students. The person introducing me commented very little on my background and expertise and focused more on my outfit and looks. Considering that the class was composed of 30 students, of which 28 were male, I was cringing inside—talk about having to build your credibility from the ground up! Anyway, I went with the flow and smiled and got on with the task of teaching.
>
> ELIZABETH CANNON, Professor, University of Calgary

Experiences that may not be comfortable or usual in your own culture are not limited to practicing women scientists. They can also occur during your training years, when, in fact, you may be more vulnerable.

> I was in Finland for an academic conference and instead of a typical North American wine and cheese reception, there was a traditional Finnish sauna for the conference attendees. In a traditional sauna, that means being naked. As one of the few women at a conference with over 50 men, I politely declined to participate, despite persistent pressure by some attendees. It is normal for saunas in Scandinavia to be coed, but at an international conference, with my male supervisor in attendance, I didn't feel the need to follow the adage: "When in Rome, do as the Romans do."
>
> NATALYA NICHOLSON, former physics graduate student

Overall, traveling to other countries and interacting with colleagues from other cultures is one of the best aspects of a career in science, as you'll read in the chapter "Training and Working abroad." It is continually fascinating and makes one think about the aspects of our own culture that have particular meanings. The key is not to immediately judge a situation that may not look familiar in your own cultural context.

When traveling to Russia, my host presented me with cut flowers when I arrived at the airport, as well as every day when he picked me up to go to the institute. One day I came off the elevator to see him standing there with a bouquet of red roses. Needless to say the meaning of red roses in Russia is not the same as in North America. I took them graciously and smiled to myself.

ELIZABETH CANNON, Professor, University of Calgary

DEVELOPING YOUR OWN PERSONAL STYLE

How, then, do you develop your personal style? First, discover what works for you, what aspects of your surface and functional styles energize you and make you feel supremely confident and ready to take on the world. One method is to keep a journal. On those days when everything seems to go right or you feel particularly energized and confident, record how you were expressing the elements of personal style discussed above. Do likewise on the days when your experience is just the opposite. You may discover some trends emerging. For example, you may feel more professional and confident in handling questions from the audience when wearing your favorite suit for an important presentation, but you hesitate to offer input at departmental meetings because you have difficulty making your points quickly and succinctly. The data you collect will enable you to repeat what works and take corrective action where necessary.

Then consider how others react to you and your style. While it may be instructive to observe their behavior and draw some conclusions, it can be dangerous to infer too much. A better approach is to seek more objective feedback from friends and colleagues. And in a male-dominated environment, the perceptions of *men* count...never underestimate the value that honest, targeted feedback from one of your trusted male colleagues can provide.

There are vast resources available for you to consult, many at low or no cost. A variety of excellent books on the various aspects of style and interpersonal dynamics are available through your local library. Clothing stores may have wardrobe consultants who will assist you in putting together a high-quality, functional, professional wardrobe within a reasonable budget. Your colleagues can provide helpful tips and feedback. Networking groups are also useful forums for observing success and exchanging information; in our workplaces (Petro-Canada and the University of Calgary), we have established collegial groups that meet regularly for seminars on a variety of topics such as wardrobe building, workplace etiquette, making effective presentations, and cross-gender communication.

Developing and cultivating a successful personal style are a worthwhile investment of your time and energy. Experiment with modifying your style gradually to find what works—but never feel you must compromise the essence of who you are in order to follow some prescribed formula for dress or behavior. And don't be afraid to evaluate your own style as you move through your career.

> Personal style is something that evolves over one's career. Taking on new positions often requires learning different ways to be effective. For example, moving to a very traditional university like Princeton hasn't changed my way of dressing or my painting in meetings, but I'm learning to be more diplomatic when suggesting that change might be worth considering. Sometimes I have worried that taking a new approach might compromise who I am. In these cases, discussing the issues with my husband, close friends, and mentors has helped me find a balance in which I believe I am maintaining my integrity while becoming more effective.
>
> MARIA KLAWE, Dean of Engineering, Princeton University

In the end, your personal style is just that: *personal*. Any advice or published information, even in this chapter, can only present guidelines. The authors themselves would be the first to acknowledge that they have never rigorously adhered to all the "rules." A competent and capable scientist who consistently produces excellent results can become very successful. But one must recognize that choosing to adopt a personal style without considering the importance of both *surface* and *functional styles* in influencing your impact on others *and* their acceptance of your work, could make that success more difficult to achieve.

REFERENCES

Bixler, S., and Nix-Rice, N. *The New Professional Image*. Avon, MA: Adams Media Corporation, 1997.

Bixler, S., and Scherrer Dugan, L. *5 Steps to Professional Presence*. Avon, MA: Adams Media Corporation, 2001.

Gray, J. *How to Get What You Want at Work: A Practical Guide for Improving Communication and Getting Results*. New York: HarperCollins, 2003.

Krupnick, C. G. Women and Men in the Classroom: Inequality and Its Remedies. In: *Memo to the Faculty* 9 (Faculty Teaching Excellence Program). Boulder, CO: University of Colorado, 1985.

McCoy, L. First Impressions. *Canadian Banker* 1996;Sept/Oct:32–26.

Molloy, J. T. *New Women's Dress for Success*. New York: Warner Books, 1996.

Morrison, T., Conaway, W. A., and Borden, G. A. *Kiss, Bow, or Shake Hands. How to Do Business in Sixty Countries*. Holbrook, MA: Adams Media Corporation, 1994.

●●

CHRISTINE SZYMANSKI, Ph.D.
Research Scientist
Institute for Biological Sciences, Pathogen Genomics
National Research Council, CANADA
Microbiologist

●●●

Dr. Christine Szymanski is at the forefront of the new science of glycomics. Internationally recognized for her studies on the intestinal pathogen *Campylobacter*, she is an exemplar: highly productive, a brilliant communicator, and a model collaborator with an infectious enthusiasm for discovery.

Her scientific career began at the University of Winnipeg (Canada), where she earned a Bachelor of Science in cooperative chemistry. She completed her Ph.D. in medical microbiology and immunology at the University of Alberta (Canada), and then studied *Campylobacter* glycobiology during a postdoctoral fellowship at the Naval Medical Research Center in Maryland (United States). Dr. Szymanski continues her work on bacterial glycobiology as a research officer at the National Research Council (Canada), collaborating extensively with scientists in academia, government, and industry and gaining a wide spectrum of experiences from basic research to commercialization and protecting intellectual property.

Dr. Szymanski has never regretted her choice to diversify her educational background because it has enabled her to use a multidisciplinary approach to solving problems and enhanced her ability to relate well to scientists with different areas of expertise. She is ideally suited to the government career path, which combines elements of academia and industry and, at the National Research Council, offers immediate access to experts from multiple disciplines who work collaboratively on common projects. As an early career scientist, Dr. Szymanski has successfully established herself. She has a vibrant research program, good publication and grant track record, strong collaborations, and a laboratory full of bright and enthusiastic students.

However involved she is in research and mentoring and promoting science to diverse audiences, she strongly believes in balancing the demanding schedule of career building with other enjoyable activities, such as sports and traveling with her husband, whom she met in graduate school. She is thankful that, as a fellow scientist, he can relate to her daily triumphs and hardships and has been flexible in moving from place to place and changing his own career directions to accommodate hers.

9

·········

COMMUNICATING

SCIENCE

·················

Christine Szymanski[1]

> *It is insight into human nature that is the key to the communicator's skill.*
> *For whereas the writer is concerned with what he puts into his writing,*
> *the communicator is concerned with what the reader gets out of it.*
> William Bernbach, the father of advertising

INTRODUCTION

Communication is an essential part of a life in science. It is about explaining the importance of your work to grant panels, funding agencies, and project reviewers; publishing your research; and informing and educating the public. It is also involved in building relationships, bringing and keeping teams together (e.g., in your laboratory, through collaborations), and becoming recognized.

[1] With contributions to "Communicating with the Lay Public and Media" by Dr. Maria Spiropulu, Physics Department, CERN (European Organization for Nuclear Research), Geneva, Switzerland (indicated by "MS").

The ability to communicate effectively is not an innate quality that one either does or does not possess. It is a set of skills that can be learned and developed, regardless of your starting point. And while some people seem more naturally adept than others, through conscious effort, practice, and constructive feedback, everyone can improve.

Many excellent books and resources exist already to help you develop expertise in the various aspects of communicating. In this chapter, we do not propose to duplicate their efforts or provide a comprehensive review. Rather, we focus on the key aspects of effective oral communication, with specific reference to presenting science to scientists, building collegial relationships, and communicating with the media and other lay public.

ASPECTS OF EFFECTIVE COMMUNICATION

The greatest problem in communication is the illusion that it has been accomplished.
George Bernard Shaw, Irish playwright and Nobel laureate for literature

Generally speaking, all forms of communication involve a **sender**, a **receiver**, and a **message** that is being transmitted. As in all dynamic systems, both sender *and* receiver require **feedback** from the other in order to optimize the "connection" between them and (ideally) enhance understanding of the message. Any change in the system will affect communication.

Regardless of how important a topic or issue may be to us as initiators ("senders"), we cannot assume that our audience ("receivers") will share our motivation and perspective. As much as we may wish it to be otherwise, we have little control over how invested others will be in any interaction with us and how attentive they will be in our attempts to communicate with them. Though we may be able to influence them to a certain extent (which will be discussed shortly), the best strategy for communicating effectively is to understand what we *do*, have control of and to use it to our best advantage: our scientific knowledge, preparedness, style, and approach; our knowledge of our audience; how we develop and deliver the message; and how we solicit and use feedback.

Be Prepared

Le hasard favorise l'esprit préparé (Chance favors the prepared mind).
Louis Pasteur, biologist and chemist

Know Your Science

All the successful scientists interviewed for this book have an intense desire to learn and discover. This motivates them to develop a thorough knowledge of their subject and read beyond their narrow specialization. Without this foundation, a meaningful exchange of ideas is not possible, especially when communicating with diverse audiences. The popular Latin quote "Scienta est potentia" (Knowledge is power), describes one consequence of knowing your subject. Another, described by Louis Pasteur in his statement, "Chance favors the prepared mind," is being able to take advantage of unforeseen opportunities when they arise. The impact on establishing key collaborations, realizing important discoveries, directing our research, and advancing our careers can be profound.

Rehearse

Being able to speak knowledgeably and fluently is important for effective communication, especially at the beginning of any conversation or presentation since it sets the tone and can positively influence self-confidence. One thing successful scientists do—regardless of how experienced—is to rehearse key points for upcoming conversations and practice their talks. For presentations in particular, practicing aloud, rather than silently, enables you to speak clearly, pronounce terms correctly, avoid stumbling through difficult terminology, adjust your pacing, and make smooth transitions between major points. Even if you do not have time to practice the whole talk—and with a busy schedule, it is often difficult to find the time to rehearse a standard 45-minute presentation over and again—it is essential to rehearse what you are going to say in the opening few slides.

Use Personal Style and Nonverbal Communication to Your Advantage

Too often we overlook the subtle aspects of communicating that can enhance or undermine our success. Regardless of our method of communication—oral, written, or electronic—how we present *ourselves* will influence how we are perceived and how the message is received. (This is discussed at length in the chapter "Personal Style.") People are more likely to take us seriously—and listen more closely to what we are saying—if we handle ourselves with professionalism and integrity and convey confidence. Ironically, even if we don't feel very confident at first, by *acting* as though we do, we may begin to *feel* confident and comfortable, and our audience will sense it.

(The chapter "Mental Toughness" suggests other strategies for enhancing confidence.) Suffice it to say that confidence in yourself and your work and an honest and sincere desire to communicate with an audience will make your initial impression a positive one and allow you to gain respect and credibility. Then, it is up to you to follow through properly, by ensuring that your nonverbal communication *reinforces* the message that you are conveying. This may be achieved using a number of strategies, including the following.

Convey enthusiasm. Enthusiasm is infectious and can inspire an audience to listen carefully, become engaged in a topic, and begin to share your interest. Feeling genuine excitement about your research is a good beginning. It can be projected into your conversations and presentations through the careful use of voice (e.g., changing your intonation), facial expressions (e.g., opening your eyes wider when conveying an interesting result, smiling), and body movements (e.g., leaning forward when making an important point).

Appear professional. Your physical appearance is very important, especially in formal meetings and presentations. It communicates much about you, even before you speak. If you are sloppily dressed, for example, people may perceive you as not really caring about personal details. This may bias their views of the quality of your research. On the other hand, you will project a very different image to an audience when you are sharply dressed. It implies a respect for your subject, your audience, and for the organization that invited you to speak. It also communicates that you are prepared, thorough, and professional.

Adopt a relaxed, upright posture. This aspect of personal style, though often trivialized, has an enormous effect on your self-confidence and sense of well-being—especially under stress—as well as on your audience's perception of you. Physiologically, sitting and standing with the body properly aligned decreases tension and fatigue. The postural muscles do not have to work as hard; full, diaphragmatic breathing is possible, and your psychological state is enhanced. (You will read in the chapter "Mental Toughness" how slow, deep breathing is a very effective strategy for moderating anxiety.)

From the audience's point of view, a relaxed, upright posture conveys energy, authority, and competence. You seem taller (regardless of your actual height), more open, and more confident (your stance is solid; your shoulders do not slouch inward in a guarded or protective manner). And because your clothes fit better, you present a more polished, professional appearance. It is important that this posture seem natural: when you appear relaxed—rather than stiff and soldier-like—your audience will perceive you to be approachable.

Minimize distracting mannerisms. If the facilities are available to you, it is extremely helpful to have at least one of your presentations videotaped. While it can be very uncomfortable to watch yourself in action, it will give you first-hand experience on your own presentations and can be most instructive. You will observe what is working well and identify what you can do to improve. For example, any mannerisms you have that can be distracting for an audience—such as obsessively pacing back and forth, frequently interjecting certain words or phrases into your talk (e.g., "OK?" and "um..."), or excessively highlighting items with your laser pointer—will become immediately apparent. During some of my first public presentations, I tended to move from screen to projector with a bounce. I was unaware of this until I looked up into the audience and observed my friends bouncing in unison in the back row! You will want to keep distracting mannerisms to a minimum.

Continually Improve Your Skills

Another important strategy for increasing your preparedness and effectiveness as a communicator is to commit yourself to the continual development of your skills. Obtaining specific feedback on your own presentations from videotapes (as mentioned above), from colleagues or from presentation specialists is one approach.

Another is to learn from role models and exemplars. For example, analyze an interesting lecture that you recently attended and identify what captivated you. I really enjoyed listening to my parasitology professor because he was so enthusiastic about the material and had an excellent style of visual presentation that always kept me on the edge of my seat, waiting to hear and learn more. My chemistry professor, on the other hand, had a very casual, yet effective, presentation style that included the use of humor and multiple examples of applications. Everyone has his or her own unique style for presentations. It is up to you to develop one that is effective and comfortable for you.

Finally, many excellent books, online resources, courses, and workshops exist to help you develop expertise in the countless aspects of communicating. Take advantage of the opportunities available to you. Be proactive. The time and effort will be well worth it.

Understand Your Audience

Another way to influence how successfully we communicate is to understand our audience—the "receivers" of the message—and to use the information to make the best "connection." Individual differences among our listeners and between them and ourselves—in education, training, experience, language, culture, temperament, style, expectations, assumptions, priorities, needs, and the like—can have a profound impact on how they "receive" our message. First and foremost, we must address the receivers' primary question: "What does this have to do with me?" Everyone has their own priorities, is working to achieve their own professional and personal goals, and is very busy balancing countless demands on their time and attention. Unless our audience immediately perceives some value in what we are communicating, they will not be engaged. Granted, it is impossible to be able to reach everyone in a particular audience, but understanding the majority and targeting the message accordingly, is the best first step. Techniques for developing a clear message that is appropriately "packaged" to enhance understanding follow.

Develop a Clear Message

> *If I am to speak ten minutes, I need a week for preparation; if fifteen minutes, three days; if half an hour, two days; if an hour, I am ready now.*
>
> Woodrow Wilson, former U.S. President

As mentioned earlier, the point of communication is to convey a message. Obviously, the better we know our subject matter, the clearer we will be on what points are essential, and the more successful we will be in creating an engaging story. Do not burden the audience with details that will obscure the bigger picture that you are trying to convey. Continually ask yourself: "What is the story I wish to tell? What is the message I want to convey?" Then stay focused on this. We cannot possibly tell it all, nor should we try. How many of us have sat through lectures or meetings in which the speaker's goal seemed to be to fill our heads with endless facts and figures? How much could we really retain? How much detail was truly necessary? We need to convey *less* rather than more information and allow the audience to ask for further detail through their questions.

Package for Understanding

No matter how well crafted our message or how learned we are, we will not communicate successfully with our audience unless they can understand what we are saying. If the message is at the right level and degree of complexity, we will be able to engage and hold their attention. To do this, we need to know our listeners. Are they colleagues in our specialized area of research? If so, they will understand the jargon of our field and will be interested in greater detail. Using too many generalizations may give the impression that we are unsure of our topic, are holding back information, or are unwilling to collaborate. Are they scientists attending a general meeting? Giving a broader picture and pointing out the applications of our work may be more appropriate. Are they visiting scholars whose first language and field are not our own? Using basic terms without scientific jargon may facilitate communication. Are they graduate students? Giving specific examples and providing applications will capture their interest. Are they venture capitalists considering investing in new technical innovations? We need to emphasize the novelty of our work and how it provides benefits not previously described. A lay audience who want to understand the science and how our topic affects them? We'll be most successful if we provide descriptions that build on a basic understanding of scientific concepts and explanations that use common metaphors.

Transmit the Message Effectively

The nature of the communication medium (e.g., face to face, written, via radio, television, Internet), will influence how the message is packaged, transmitted, and received. The fact that we may be effective speakers, for example, does not imply that we are equally effective writers, so we need to be skilled in using multiple media and be flexible and adaptable in our approaches. Regardless, there are some key techniques—common to all media—that are important to employ: repetition, emphasis on significance, pointing out novelty, providing applications, and giving examples. In addition, the following strategies may help you to improve your "connection" with your listeners in communicating orally.

Convey Confidence

This topic cannot be overstressed. We've already discussed how preparedness and nonverbal communication (if used effectively) can enhance confidence; it is equally important *during* communication to convey confidence. One technique to decrease anxiety is to use the strategies of mental toughness (discussed more fully in another chapter). It is normal to be anxious about speaking in public. This is particularly true in situations in which you are being examined for a new position or promotion. Ph.D. students, for example, always feel very nervous when they go into their comprehensives or thesis defenses because they will be examined by an expert panel. It is important to remember that you are defending work that you have concentrated on for a considerable period of time, and, in this respect, it is *you* who really is the person most familiar with the topic.

No matter who your audience, there usually is some piece of information that you can contribute to the conversation that will be novel or interesting. Think of it in this way: your audience is there to learn from you, not just to criticize or humble you. They are interested in gathering new information and wish to discover how much you have learned. When viewed from this perspective, the situation does not seem quite as intimidating.

Use Appropriate Language and Vocabulary

When speaking, it is important for people to be able to hear what you are saying and understand the vocabulary that you are using. Articulate the words clearly (especially technical and scientific terminology) and project your voice so that the entire audience can hear. Use a microphone if necessary (and available). To prevent a monotone delivery, modulate your voice. Maintain good posture, even when you are not physically in the presence of your audience (e.g., during a radio interview), for it will positively affect your voice and the nonverbal messages that you convey through your speech. Just as the diaphragm is the organ of breathing, it also is the power behind your voice. Your audience does not have to see your body to hear your enthusiasm and confidence. Remember also to pause, so your listeners have time to process your message.

When describing complex ideas or processes, help people visualize things a little differently through the use of analogy. Even with very technical, specialized knowledge and vocabulary, it is possible to simplify concepts and use common language to make your points. Imagine, for example, that you wish to describe Le Chatelier's principle, which states that when a chemical reaction in equilibrium is subjected to a stress, the system will adjust to relieve the stress and restore equilibrium. You could ask your audience to visualize a flat packet of ketchup and them squeezing the packet at one end: the ketchup will be forced to the other end. This is a simple analogy of how chemical reactions move back and forth between reactants and products.

Personalize and Use Humor Appropriately

Adding an appropriate personal touch to your message can be an effective way of engaging your audience, changing the pace, and maintaining interest. For example, in a talk given by a veterinarian advocating the use of animal research in science, the speaker showed a slide of himself together with his family and their pets. This image effectively demonstrated the point that he was a person who was passionate about animals who also realized the importance of using animals in research. In another, a pediatrician with interest in bacteria infecting the human intestinal tract spoke of the considerable advances in the field. He showed a slide of his children reading *The Journal of Gastroenterology* and joked about how they were keeping him abreast of the literature. Many scientists also show photos of the people who have done the work in their laboratories rather than provide a long list of names. This gives the laboratory personnel more exposure, especially at conferences where they may be subsequently recognized—for example, at their posters or in other sessions—and approached with questions about the work. The author works with a bacterium called *Campylobacter jejuni* that normally lives within the intestinal tract of chickens, so she likes to superimpose the heads of her students and collaborators on an image of a chicken flock for her final acknowledgment slide.

Another approach to using humor is through cartoons or jokes. These do not have to be complex—in fact, simpler may be better—they just have to make a connection with the audience. However, be cautious. Humor can be both very personal and very specific to one's culture, so it has to be appropriate. Judge your audience carefully. When in doubt, it would be best not to use humor. One scientist who gave a talk in another country tried to include humor by joking about his hosts' culture. Because he was foreign to the country, the audience gave him some latitude, but many people approached him after the talk and said that his humor had been offensive.

Be Comfortable with Silences and Solicit Feedback

Become comfortable with silences. Often, when we feel uncomfortable in silences, we tend to talk on and on to fill the void. This does not enhance communication and often makes us appear as though we lack confidence. Instead, silences may offer a good opportunity to solicit feedback. After all, communication is an *interactive* process. Asking your listeners questions gives you an opportunity to find out whether you have conveyed your message clearly (so you can address points of confusion), determine what their areas of interest are (for further elaboration), and also learn from them (e.g., different perspectives, interpretations, techniques). Each experience provides you with more opportunities and chances for development—as a scientist and as a communicator.

PRESENTING SCIENCE TO SCIENTISTS

The most important part of science is reporting results in a concise and convincing manner. Without the communication of results, even the most brilliant ideas would be lost and unrecognized.

Much of our communication as scientists involves sharing our results with other scientists, through articles in journals, presentations at meetings, invited seminars, and informal conversations with colleagues. The focus of this section is on methods for communicating effectively through oral presentations and poster sessions. Also discussed are how to deal with questions and manage conversations during these sessions.

Oral Presentations

Applying all the elements of effective communication (described above) will help you deliver an excellent oral presentation. You need to determine what you want to say and say it clearly. When communicating with people (in presentations and conversations), be very clear about your objectives, emphasize them, and remember to modulate your voice and convey enthusiasm. Obviously, it is important for people to be able to hear what you are saying and to understand the words that you are using. So face your audience (rather than the board or screen), project your voice, and pronounce each word clearly (especially technical and scientific terminology). Also practice your talk and ask for feedback from your mentors and colleagues.

Define Your Objectives

Having well-defined objectives is an important strategy for creating effective presentations. If you have 45 minutes in which to speak, you cannot possibly cover everything. So you have to decide what your final, "take-home" message is and what points you would really like to convey. These can be emphasized throughout the talk and then summarized as a few bullets during the conclusion. As a scientist working in a particular area, every single detail is very important to you because you have worked very hard to prove these details. But if you overwhelm your audience with too many facts and figures, they will lose interest and stop listening. On the other hand, if you emphasize the relevance of the work and summarize the key experiments that were done to prove your point, there will be a higher probability that your listeners will remember the objectives of your presentation.

Add Color and Images to Slides

Adding some color to your slides and summarizing key items in pictures (rather than using many words in bulleted points) will add interest and convey your points in multiple ways. Your presentation will also appear more professional and will convey to the audience that you put in that added effort. Avoid using sentences in presentation slides because this will encourage the audience to read rather than listen to what you have to say. Images will leave a lasting impression with your audience and help them understand complex theories much more readily.

Deliver Material at the Appropriate Level of Complexity

The importance of targeting your audience appropriately, though discussed earlier, cannot be stressed too much. The vocabulary that you use and the degree of detail and level of complexity that you present should depend on your audience. With homogeneous groups, you can assume a higher level of understanding and can use more of the jargon associated with that field because people will understand what you're saying and will expect you to "speak their language." If, on the other hand, you are addressing an audience from diverse scientific and technical backgrounds, it is important to describe things in basic terms, explaining details clearly and using analogies that will be familiar to most people. This applies if you are addressing interdisciplinary teams, broad scientific conferences, the media, or the general public. I often give presentations to colleagues from many disciplines. When speaking about concepts that span a range of scientific fields, I describe them in general terms. By aiming for a basic level, I am confident that my message will reach everyone; during the question period, I can focus on discipline-specific details and nuances.

Effectively Pace Yourself

It is quite normal to speak quickly when giving a presentation. Many people do so because they are feeling nervous, while others try to relay too much information and speak quickly to cover everything. Some are so enthusiastic about their research that it bubbles over into their delivery. It is important to deliver the talk at a pace that keeps the listeners engaged and able to follow the reasoning.

To avoid problems, you may wish to scan the audience while you are speaking and make eye contact periodically. This can give you feedback on their level of engagement and comprehension. We described earlier the importance of using nonverbal cues to communicate; your audience is doing the same. Look to their body language for feedback. There are certain signs that may be indicators that your audience is confused or inattentive. If, for example, you see people furrowing their brows or looking puzzled, then you will know to slow down and attempt to explain things in more detail. On the other hand, if most people are nodding in agreement, shifting in their seats to move closer, or taking notes, you can assume that they are following your line of reasoning and the pace is satisfactory.

Obviously, there are many reasons why audience members may be frowning, yawning, or shifting listlessly. They may, for example, be preoccupied by something completely unrelated to your presentation, be tired and fighting to stay awake, or suffering from lower back pain and are uncomfortable. However, if too many people are exhibiting the same behaviors, you may assume that one reason might be the pace or content of your talk.

Some people are uncomfortable looking at the audience because they are too sensitive to the listeners' reactions and inappropriately misinterpret the behaviors as being directly related to their presentation. This may undermine their confidence and cause them to panic during their talk. For these presenters, it is less advisable to look directly at your audience; you may wish to scan just above their heads, looking only for signs of nodding. What you choose to do will depend on your own comfort and confidence level and can be modified as your presentation style develops.

Handle Questions with Confidence

Think of questions from the audience as opportunities to test your current understanding and interpretation. You need not lose confidence if you do not know the answers to all the questions that are posed, for you cannot possibly remember all the details or have thought of everything related to your presentation. In fact, the questions themselves may help you to refine, for example, your methodology and analyses, to give your work a more solid foundation.

You may wish to bring copies of the key papers relating to your topic to the presentation (if you highlighted the important points during your review of the literature, all the better). In this manner, you do not have to remember every detail of every study you are referring to and you will have the information at your fingertips. If a member of the audience is interested in some nuance, you will be able to refer them to the appropriate article and provide a copy for them to look at. The gesture will be most appreciated and actually demonstrates that you know your material and refer often to the literature.

Another way to handle questions is to suggest possibilities based on similar models, systems, or organisms. For example, if the answer to a question asked about your presentation is unknown, but you are aware of a similar mechanism used by another organism, you could speculate that a similar process occurs in your system. By this method, you are still demonstrating your knowledge of the field, even if you do not know the particular answer as it relates to your topic. Another strategy is to have a trusted colleague listen while you rehearse your talk and ask obvious questions. You may then be able to anticipate what questions will be asked and prepare the appropriate answers. If you have the time and are skilled at prediction, you may even prepare some slides that would address these topics.

When answering questions that seem simple or suggest that the person was not listening, always treat the audience member politely and with respect. Give them the benefit of doubt and remember that it is sometimes difficult to be fully alert during an entire presentation. You do not want to address the individual in such a way as to make him or her feel that the question is stupid or unreasonable. And it may be beneficial to the audience in general to hear you describe the point again—in a different way and without making reference to the fact that it was already covered—for others in the audience may also have missed it. Regardless of the question, you certainly want to say something; you do not want to respond with simply a "yes," "no," or "I don't know." This conveys a certain lack of professionalism and gives the impression that you are not even willing to venture an intelligent guess or engage in conversation.

Remain professional when dealing with aggressive or abusive questioners. Regardless of how others behave toward you, always maintain your professionalism when communicating with people. If you feel attacked or if the questioner seems to have a hidden agenda or is being abusive, remember that others in the audience are just as likely to think so too. Choose not to take the delivery personally; detach yourself from any negative emotions associated with the question and treat the inquiry as legitimate. (See the chapter "Mental Toughness" for more strategies for not taking things personally.) Always leave them with the knowledge that *you* are the one with information to share.

Conversations with Others: What to Share, What Not to Share?

Obviously, when communicating with other scientists, it is important to share your knowledge, but be protective of your unpublished work and cautious about how much you reveal. Knowing what to say and what not to say develops with experience—and depends on your area of study and your field of employment—but, as a general rule, it is usually best not to disclose too much information about anything that is not ready to be published. This is a good topic to address with a supervisor or mentor before attending a meeting. You can practice what you plan to present, and he or she can point out obvious questions that will be asked and topics that should not be described in detail.

> During one of the first meetings that I attended, I was approached by a leading scientist in my field of study. As an eager graduate student, I was anxious to share everything, so I described all the projects that I was working on—in detail. When I asked him what he was working on, he simply said, "similar things as before." I learned a valuable lesson that day: most people do not describe unpublished data in any detail because science can be competitive.

Confidentiality is also becoming more and more important now as the scientific community moves toward commercialization. Many organizations outside of industry are being pushed to protect intellectual property, so certain aspects of your work may not be allowed to enter the public domain (i.e., be disclosed publicly) until the work is protected. It is advisable for the inexperienced scientist to be coached by someone more senior. This would also enable him or her to prepare and rehearse specific phrases that can be used to address questions that cannot be answered tactfully in any other way. For example, the phrase *"I am sorry, but I can not talk about those results right now because they are in the patent process"* would be a respectful response that would not divulge confidential information.

In contrast, remember not to flood people with too much information. Try to have your defined objectives in mind so you can emphasize these. For example, when conducting a tour of our organization to a group of students, I presented several statistics on *Campylobacter jejuni*, demonstrating the significance of this bacterium as a food- and water-borne pathogen, describing why the National Research Council studies this organism, and relating the relevance of the studies not only in Canada, but worldwide. At the end of my session, I told them that they probably would not remember many of the details that were presented but should at least remember the name *Campylobacter jejuni*. (I knew they hear about *E. coli* all the time, but thought that if they remember the name *Campylobacter jejuni*, they would now know the name of North America's most common food-borne pathogen.) At the end of the day, our communications representative prepared a question for the students, along with a prize. The question was "What organism was responsible for at least half of the reported illnesses during the Walkerton outbreak?" This exercise tested whether my message came across and it did!

Using Feedback Effectively

The questions that you receive from your audience are a direct indication of your success in targeting the audience, packaging the message, and presenting the information. If, for example, many people ask for clarification of what you thought were rather basic concepts, you can conclude that your assumptions about the knowledge base of your audience were incorrect and that your material was too complex. This kind of feedback allows you to take a step back and explain again the key concepts required to understand your main message. It is never too late to supplement your presentation with new information because your lecture is not over until you have left the room. You can also design the presentation to require feedback from the audience (as often occurs in teaching and workshops), by asking specific questions during your talk. This will enable you to fine-tune your delivery as you go and gather information about your audience that will allow you to make a better connection.

As you become more confident in making presentations, you may be able to include additional last-minute comments as information becomes available. For example, if the speaker before you mentions a point that is also in your presentation, you can—in your own presentation—demonstrate that you were attentive to their message and reinforce the point, by saying, "as you heard in the previous talk..." However, adding comments to your presentation in this way can only be done effectively when you have considerable experience and a high comfort level.

Poster Sessions

As a young scientist, presenting your work at poster sessions is just as important as giving an oral presentation and provides similar opportunities. Practicing the manner in which you plan to present the information in your poster—before leaving for the meeting—can be very helpful. Even though you have been actively involved in assembling the poster, the manner in which you describe the data will be an abbreviated version of what is actually written down. You need to be prepared to summarize the key points concisely and answer questions related to the topic, for at the conference, people will often stop at your poster and, instead of reading everything, will say, "Walk me through your work."

It is also important for new presenters to convey confidence during this process because poster sessions will be one of their first opportunities to meet senior scientists and build future relationships. Making a good first impression and being able to describe the work with confidence and enthusiasm can advance your career in unexpected ways (see the chapters "Personal Style" and "Networking" for more information). Take note of the names of the people who approach you—you may even recognize the names of key people in your field of study, which will provide you with an additional opportunity to say "Hello, Dr. X. I've wanted to meet you...," or "I've been interested in this aspect of your work..." This can leave a lasting impression with that person because you have made a personal connection with them.

BUILDING COLLEGIAL RELATIONSHIPS

Communication is about making connections with people. If you can not build collegial relationships effectively, you are not likely to succeed.

In this section, you will learn the importance of communicating effectively to create relationships with people with whom you have something in common. These connections may help you build a network, establish mentoring relationships (with you as mentee or mentor), and form collaborations. The focus of this discussion is on the role of personal style, identifying mutual interests, and the use of electronic communication in the development and maintenance of collegial relationships, and on having clear priorities and managing interpersonal conflict. (You may wish to refer to the chapters "Networking" and "Mentoring" for more detail on each of these topics.)

The Importance of Personal Style

When developing and maintaining relationships, what we communicate *nonverbally* can be just as important as what we say. By becoming aware of our personal style and how it can enhance or undermine our verbal communication, we may become more effective in establishing successful collaborations and working relationships. Though an entire chapter is dedicated to "personal style," we briefly revisit it here.

Honesty, Sincerity, Humility

Honesty and sincerity are very important in science—in our research activities, interpersonal relationships, and communication. We do not often think about these aspects of personal style, but they affect how others perceive us and our work, and they are critical in negotiation and conflict resolution.

Also important in interpersonal relationships is not to appear to be more than you are, nor pretend to know more than you do. As knowledgeable as you may be in any given subject area, there always will be people who have more knowledge and experience. So have confidence in what you *do* know, be humble in your ignorance, and maintain a willingness to learn. This way, you will earn respect.

Be an Active Listener

Most conversations are simply monologues delivered in the presence of a witness.

Margaret Miller, American writer

To be an effective communicator, you need to develop active listening skills. By observing the body language of the speaker, listening attentively, focusing on understanding what he or she is trying to convey, and rephrasing what he or she has said, you will be sure that you have understood the message. The latter point is a very constructive technique of effective communication because, by summarizing (in your own words) what a person has said, you clearly demonstrate to them what you have understood (i.e., you are giving them feedback) and you will also receive confirmation that you have received the message correctly.

Other people prefer to take notes because they have a better visual memory or find it difficult to remain focused for longer periods of time and cannot process the information simply by listening. Note taking also allows you to reinforce what you have heard and occasionally provides you with more resource material at your fingertips during subsequent discussions. Remember that *listening* and accurately *receiving* the message are as important to the communication process as are *formulating* and *sending* the message. As you are listening, use body language to indicate that you are following the line of reasoning and understand (these are the very signs that you would look for when speaking), e.g., occasionally nod, smile, use other facial expressions to convey interest and comprehension.

Identify Mutual Interests

Tell me and I'll forget. Show me and I'll remember. Involve me and I'll understand.
                                        Confucius, Chinese philosopher and reformer

Usually when you are invited to give a talk at an organization or appear for a job interview, you spend the day with the employees and meet with some individually. Rather than going in blindly to speak with them, do some research before your trip to discover their interests and where there might be possible overlap with your own field of study. By doing your "homework," you will be able to stimulate conversation, demonstrate that you are interested in what they do, and have some questions and ideas about the topic. They, in turn, will see you as an interested colleague who may also benefit them and the organization.

Preparing for such visits is very simple to do. Occasionally, the organization that you are planning to visit will provide you with an itinerary ahead of time, from which you can obtain the names of the people you'll be meeting. Most groups have a Web site that lists individual employees, along with their interests and publications. But, even without an itinerary, you can visit the institute's home page and familiarize yourself with the mandate, areas of expertise, etc. Try to incorporate some of these interests into your talk, as the following scientist did:

> Recently, I gave a presentation on bacterial sugars to members of a parasitology department. In order to connect with the audience, I did some background reading on the importance of sugars in parasite biology and based my introduction on a summary of those points. By using that introduction as the starting point for my talk, I was able to engage the audience. It was clear to me that they were really interested in hearing about my work on the roles sugars have in bacteria because of the possible application it may have to their own studies. An additional benefit was this: in browsing through the parasitology literature for reports on sugars, I was exposed to studies that I would not otherwise have come across. So, by putting in that little bit of extra effort, I increased my knowledge base, encountered different perspectives from the literature and the discussions following the talk, and doubly impressed my audience by being prepared for the visit and by my knowledge of the topic. This also increased my opportunities for future collaborations and networking.

Communicating Electronically

> *The newest computer can merely compound, at speed, the oldest problem in the relations between human beings, and in the end the communicator will be confronted with the old problem, of what to say and how to say it.*
>
> Edward R. Murrow, father of broadcast journalism

A new form of communication that continues to grow in popularity is electronic mail. There are several approaches to e-mail communication (as there are with letter writing), from informal greetings to more formal messages that provide updates to collaborators or accompany applications for employment. Similar to any form of communication, you do not want to overburden the recipient with long messages. Too many people send e-mails that go on and on and overwhelm the receiver. It is difficult to read such letters on the screen and inconvenient to have to print the message in order to digest it properly. When composing e-mail messages, therefore, it is a good idea to break up ideas into separate sentences or sections. Some people use different text colors or highlight sections in bold. Make sure that the subject line states exactly what the message is about—similar to a title in a presentation. Generally speaking, make your e-mail message concise. Do keep in mind for whom the message is intended; you may, for example, wish to balance the message with a short personal update if the recipient is a close friend.

E-mail messages can also be an effective way of distributing agendas before meetings or when you are planning a long conversation over the telephone. As discussed in the chapter "Time Stress," having written agendas for meetings—however informal—is a good strategy for keeping the conversations focused. They also make it easier to record important notes. E-mail enables you to communicate with a larger group of people by using distribution lists and copying multiple recipients on the message. Putting technology to work for you in this way can reduce the number of your face-to-face meetings and conference calls.

E-mail Etiquette

People are very busy and have very little time to review their e-mail. It is a good idea, therefore, to take the time to proofread your message to make sure that what you want to say is clear and cannot be taken in the wrong context. Unlike in a conversation, the receiver cannot quickly ask for clarification if something is unclear, and the sender cannot make a correction once the letter has been sent. Sometimes you will write down an idea and after you have had a chance to think about it, you realize that the question being asked does not make sense. So, it is occasionally useful to leave a message in your draft box, until you have had the opportunity to think about it more clearly. This will decrease the amount of e-mail traffic and the need for clarifications.

Be as professional in your electronic communication as you are in person or over the telephone. Remember that if you are writing to someone for the first time, your e-mail message will be making your first—and often lasting—impression on that person. It is also important to know whom you are addressing; your e-mail needs to reflect the appropriate tone of respect and appreciation. Always pay attention to proper grammar, sentence structure, spelling, and punctuation. Avoid writing in a chat-session format.

I recently received two e-mails that should have been checked before they were sent to me. In one, the person was asking whether I had any positions available in my laboratory, but the sender had not proofread the letter to notice that the name of the person being addressed had not been changed. This was a clear indication to me that the sender's interest in working in my laboratory was not sincere and I did not reply to the message. In another e-mail message, the writer began the letter by praising a recent paper that I had published, but then asked me whether I had thought about doing a certain experiment that was already described in that same paper. Clearly, the person had not actually read the paper and, again, was not sincerely interested in my work.

If used properly, e-mail can be an efficient and effective method of communication—for initiating, as well as maintaining relationships.

Networking

Science is political. In order to have a successful career in an increasingly multidisciplinary world, you need to work carefully and diligently on projects that interest you *and* others (e.g., granting agencies, journal editors, other scientists) and invest time in building relationships with your scientific colleagues, i.e., you must be a team player. To develop an international reputation and increase the potential for collaboration, you need to establish not only a critical mass of people working with and for you, but also a critical mass of contacts (i.e., a strong, broad network). By attending different meetings, publishing and presenting your work, and participating in social functions, you will become better known, your contacts will increase, and, with them, your potential sphere of influence. You will make a greater impression on people when they meet you in person and see you in action. If they are impressed by your dedication and the quality of your work, they will be more inclined to suggest your name to other colleagues and you will attract even more collaborators. (The chapter "Networking" explores this topic in greater detail.)

As a young scientist, it is normal to feel uncomfortable about speaking with new people at meetings, especially when they are more senior. It can be daunting because you do not want to embarrass yourself. However, be encouraged. The effort is worth it. And with a good sense of humor, you can overcome your discomfort.

> My first encounter with a senior scientist whose work I had been following during my Ph.D. program was rather amusing. I went to an American Society for Microbiology meeting and was talking to a colleague at her poster when I asked if she knew whether a certain senior scientist was attending the meeting. I mentioned that I would really like to meet him because I had a great admiration for his work. Then as I looked over my shoulder, I noticed a man reading her poster. He said "hello" and lowered his notepad from his name tag—it was the very scientist I had been speaking of! Instead of being embarrassed, I commented on how happy I was to finally meet him. This is how we became friends.

Networking at meetings allows you to share ideas and return to the laboratory feeling invigorated, with fresh thoughts, different perspectives, and new experiments to try. When you go to these conferences, you often travel to very nice locations with a group of people with similar interests to your own. How often are you in a situation such as this? Take full advantage of the opportunities by attending the planned social activities as well as interacting with people during the poster sessions or after they have given their talks. This often is the best time to become acquainted. And by attending the various mixers, you will have the opportunity to meet different people and, possibly, form long-lasting friendships during your scientific career.

You may also choose to attend smaller meetings where it can be easier to connect with people. At the larger conferences, there may be simultaneous presentations preventing you from meeting all the people in whom you may be interested. At smaller conferences, participants normally move as a group, so the same people attend the same talks. Often the lunches, dinners, and social activities are also combined, so by the end of the meeting, you have become familiar with everyone and their interests. Smaller meetings may also be less overwhelming for the introverts because there is continuity and less initiative required for interaction. And it is always worth your time or effort to market yourself, your science, and your place of employment, regardless of the audience size or composition.

Establish Clear Priorities When Maintaining Relationships

As your scientific world expands, you will need to balance the requirements of establishing your career with networking and maintaining collaborations. It is not possible to keep in touch with everyone you meet, nor is it necessary or even reasonable. Some people will reconnect periodically (e.g., at conferences), while others will keep in touch on a regular basis. The frequency and level of communication depend on the individuals involved. Though e-mail has become an efficient way of communicating, it too can become burdensome. This is why it is most important to be clear about your priorities and follow through first on those that are most important. If, for example, you are working with someone to submit a manuscript at the end of the week, you will be in touch with him or her every day and may not have time to respond to your other e-mail messages immediately. With experience, your anxiety about accumulating e-mails will decrease because you will learn that other people are equally busy, have deadlines too, and will understand if your response is delayed. However, it is nice to let people know that you are thinking of them from time to time. For example, some people enjoy sending a short greeting (by e-mail) to their collaborators for special holidays. Occasional informal communications such as this provide opportunities to keep in touch and remind your colleagues of your connection.

Say "No" Diplomatically

You will continue to meet people and develop new contacts and collaborations throughout your career. Should you concern yourself about overextending yourself and your resources? The answer is "yes." At some point you will have established the critical collaborations that are mutually beneficial and realize that your group cannot expand any further. Be aware of your capabilities and limits. If, for example, a person contacts you to request a published strain, you are obliged to send it. However, if you are approached to do studies that would be too demanding on your personnel and equipment, you may decline gracefully. Send an honest reply mentioning that the studies that they propose are very interesting and explain that you currently lack the resources to work jointly with them. Or, someone may approach you to do experiments that either you or another collaborator are currently doing; you would need to mention that participating in the proposed studies would be a conflict of interest. The key is to be diplomatic. Many people simply ignore e-mail messages and do not respond at all, but a short, truthful reply is more constructive because through it, you will be treating everyone with dignity and respect.

Managing Interpersonal Conflict

Everyone will respond differently when exposed to conflict. The key is not to react mindlessly, but to pause...step back and count to 10...and compose yourself so that you can respond appropriately. It is important—always—to remain professional. Avoid becoming emotional in a situation, even if it appears unfair. Remain focused on the issue or point of contention—rather than on how you are feeling at the moment—and on how to resolve the conflict. Do not allow yourself to be manipulated or rushed into a decision that requires more time to consider. Scientists at the beginning of their career may be eager to advance quickly by avoiding conflict or acquiescing, but it is important to patiently, yet persistently, stay true to what you believe. (For a more detailed discussion of how to develop "mental toughness," see the chapter dedicated to this topic.)

Often, resolving conflict involves making difficult choices. You may, for example, be faced with a decision that is against your beliefs, but is—according to your own analysis—the best thing to do. Sometimes it is about choosing the lesser of two evils. Keep in mind that you are still moving up the career ladder and that, as a more junior member of the organization, you will not have access to all the information that is available to those who are more senior. Even so, you will not agree with all their decisions. Take the initiative to learn what mechanisms exist for resolving conflicts within your context and use them appropriately.

Whenever there is a conflict in our organization, we have several meetings and discussions on the topic to allow everyone an opportunity to speak to all the issues. Occasionally, some people who feel uncomfortable voicing their opinions in a large group would approach me on their own, outside of the meeting. The group would then meet again and I would summarize everyone's opinions by beginning, "Some of you might be feeling this way...," and not mention names specifically, but bring all comments and perspectives to the table. This way, everyone becomes aware of all the concerns, and discussions of each can result.

It is important to remember that conflict may arise as a result of individual differences in values (e.g., work ethic, beliefs, cultural norms and expectations, gender, language, personality, temperament, etc.) These differences need to be recognized, appreciated, and respected. If you view diverse perspectives and opinions as being complementary—rather than divisive—factors, they will strengthen, rather than weaken, the group.

One fact that must be accepted is that, try as you might, *you simply cannot get along with everyone,* all the time. But be careful not to destroy any future interactions. Keep in mind that you spend a lot of time choosing your spouse—whom you see for a few hours a day—while you interact with your coworkers for a longer period of time each day, and these are not people whom you have chosen to be with. So you will be more compatible with some people than others. Slowly you will learn whom you work well with and will form a really solid scientific group. This is not to say you should ignore the people who do not complement your current research directions, just that you need not invest enormous amounts of time and energy in relationships with them.

Conflict Consumes Emotional, Mental and Physical Energy, Time, and Creativity

You probably do not even realize how much your thinking processes are affected by emotional problems, such as not getting along with a particular coworker. Attending meetings to resolve conflict and trying to deal with everyone's different perspectives becomes draining on many levels: emotional, mental, and physical, as well as in terms of your time and creative energy. So it certainly is in your best interests to remain as balanced as possible and not to let things bother you. Remember to pause and put things into perspective. Let conflict be a positive motivator in your life. Naturally, you will be discouraged along the way, this is part of life, but perseverance will see you through. Decide on your priorities and what battles are really worth fighting. (The chapter "Time Stress" discusses priority setting in detail.) Choose to learn from conflict. Say to yourself, "I will take what positive things I can from this, instead of obsessing about how unfair the situation was." Focus more energy on how to avoid similar conflicts in the future by changing you policies. Things will never be perfect, but you *can stay focused on your research* and *stay focused on having the best experience possible*. By always looking for methods to make the *whole system better* and communicating these ideas to your colleagues, you may succeed. It is all about communication, negotiation, and patient persistence.

COMMUNICATING WITH THE LAY PUBLIC AND MEDIA

Introduction[MS]

When I was asked to contribute to this chapter on communicating science, I immediately thought of a comment made by French-American historian Jacques Barzun, in his *Science: The Glorious Entertainment* (1964) that "art is life and science is the antithesis of life." I have struggled for many years to comprehend the meaning of this statement and have (secretly) been slightly offended by it. Surely he does not mean what he seems to be saying. Artistic expression is probably one of the most fundamental attributes of humans and a means of communication in itself. In fact, there are manifold ways that artists express themselves, usually defined by the art form itself: an exhibition, concert, theatrical performance, dance, book. Art is easily sharable because it "speaks" directly to our emotions, our sensations, and our thoughts. In this respect, science is different. It took a long time for science to become easily disseminated and communicated, even among scientists themselves, and it does not engage our emotions or sensations in the same way as art. It is the enormous difference in experiencing and communicating art—compared with science—that presents this apparent schism: assuming that communication is a basic characteristic of life within human society, it is this lack of scientific communication that is the "antithesis of life." This has serious implications for society, for only the contribution of research-based knowledge to the discussion of some of the challenges facing the world and the effective communication of scientific discovery—to the lay public and policymakers—will enable science to influence the development of sound public policy.

[MS] Contributed by Dr. Maria Spiropulu, Physics Department, CERN (European Organization for Nuclear Research), Geneva, Switzerland.

Speaking with Nonscientists About Your Work[MS]

Scientists have a tremendous advantage in exchanges with lay audiences: people around the world are more than interested in scientific questions and their answers, in the status of the current research in pretty much every field and discipline; there is great fascination, and people are really drawn to discussions with the experts. However, communicating effectively about our work is a skill that requires commitment and practice. Lay audiences do not have the benefit of our training, nor our knowledge of the specialized language that we use when communicating with each other. Further, their interests often tend toward the practical, the common-sense applications of our word rather than the theoretical: "So, what's in it for society, really?" seems to be their main focus. This is not to suggest that their questions and interests are unimportant, rather, that we need to *understand* our audience, in order to establish a connection with them that will facilitate the sharing of information and ideas.

Further, as scientists, we are not generally compelled to emerge from our preoccupation with our own research to make what we are doing accessible to everyone, not even to scientists from different fields. There is a degree of subject specialization and descriptive language that can hamper understanding. There is little immediate and personal reward in investing time—and it does require a lot of time to do a good job in talking about what you *know* (teachers of science at every level are witnesses to that), let alone about what you don't know but are trying to discover. And it is even more difficult to talk about it in a nonpedagogical way.

> I remember speaking with a journalist once who apologized for his lack of knowledge of a subject before our interview. His enthusiasm and his willingness to be a mediator and inform people were very clear. He said to me "I don't want you to teach me, I want you to tell me about it."
>
> In the beginning of my career, when I took time to speak with people about my work, the urge to teach them was uncontrollable. But all didactic methods I knew and employed were ineffective. I lost the listener in a chaotic meandering of manifold details, and in the end, I lost the threads myself and found it difficult to make sense of my own sentences. After many years of communicating science to lay audiences and the media, I learned how to communicate effectively.

We need to do a lot of *explaining*. We need to make sense of things that do not belong in the realm of common sense to illustrate otherwise intangible attributes and notions that are perceived as preposterous. The

[MS] Contributed by Dr. Maria Spiropulu, Physics Department, CERN (European Organization for Nuclear Research), Geneva, Switzerland.

variables and descriptions that we use to describe them often have little to do with what we experience in our everyday life—quantum mechanics is one vivid example. So we need to use analogies and metaphors that will have meaning for our audiences. Even so, we won't be able to come close to satisfying the curiosity of some questioners who ask "What is it, exactly, that happens?" When this question arises, and it always does, presenting the natural phenomena as they occur, and as we observe and measure them, seems to ease people's skepticism and usually generates another level of questions. There will be, of course, cases in which our research is ongoing and we don't know the answers. Telling the audience just this can generate a richer appreciation for the wonder of Nature that scientists just love to inspire. You will find—perhaps to your surprise—that people do grasp (and forgive) that in the sciences (fundamental sciences in particular), we have approximate answers to the approximate questions that we pose, and we only uncover the mystery one step at a time.

Speaking with the Media

Understand that when you receive a query from the media, it is necessary to respond—and respond quickly—if you wish to disseminate information about your work, or address current issues that relate to your area of expertise. People in media work under tremendous time pressure and their deadlines do not allow for any hesitation on your part.[MS]

It is important to control the interview as much as possible. Though the interviewer may have done his or her homework, he or she may not know the topic well enough to ask the best questions. So make it your responsibility, in preparing for the meeting, to know what points need to be made. And in the interview, be sure you make them. Also before the interview, rehearse the few important, key phrases that relate to your work and the message you wish to convey, so that they come out naturally. That way you know what they sound like and your voice will not have a nervous squeak during the conversation. Interviewers can take you down different routes and different areas of questioning. But you are there for a purpose and want to emphasize a few certain points, so you definitely want to practice these out loud.

In your replies to their questions, be short, clear, and vivid and use language that is not too technical. But be cautious about oversimplifying things in an effort to avoid a highly convoluted answer. It is quite acceptable to say to your interviewer that the particular subject cannot be explained in simple terms and you need time to think about it and formu-

---

[MS] Contributed by Dr. Maria Spiropulu, Physics Department, CERN (European Organization for Nuclear Research), Geneva, Switzerland.

late a coherent response. In most of the interviews that I have given, there were several iterations of questions and answers before the journalist was satisfied with both. This was especially true of interviews for articles in science-related print media (e.g., science columns in newspapers, special scientific issues of news magazines, popular science magazines). Be patient with the process and creative with your use of analogies and metaphors when explaining your topic. Think, for example, of how would you explain a scientific concept, an experiment, or the results of your data analysis to your parent or to a nonscientist friend in a coffee shop. What kind of image would convey your meaning as accurately as possible without making it obscure? Try it out, and if the interviewer doesn't understand, make adjustments, try another image, ask him or her to clarify where he or she is having difficulty. The interview is, after all, a conversation, not a lecture. (Recall comments made in the previous section—about *explaining* rather than teaching.)

Sometimes, the interviewer may have a hidden agenda and begin asking questions on topics unrelated to your expertise. In situations such as these, it is important to know how to respond to the questions and redirect the conversation to topics about which you are more knowledgeable and comfortable. Or, you could refer the journalist to an expert on the topic (they both may be grateful). Unless you really are sure of what you are talking about, do not respond to such questions. Otherwise you could end up making statements that are incorrect, appearing very unknowledgeable, and feeling panicked and embarrassed. A high degree of diplomacy and professionalism is needed; you are, after all, representing your organization and scientists in general.

> A situation such as this came up when I was being interviewed live on television. The discussion was supposed to be about the E. coli/ Campylobacter outbreak in Walkerton, Ontario, Canada,[2] but the interview was conducted during the time that the severe acute respiratory syndrome (SARS) was in the news, and the journalist was interested in my opinions about SARS. I replied by redirecting the interview to my field of study: "Canadians are concerned about infectious diseases and that is why scientists at the National Research Council are working on Canada's number one food- and water-borne pathogen, Campylobacter jejuni, which was responsible for nearly half of the cases reported during the Walkerton outbreak." In this way, I was able to return the conversation to my area of expertise.

Also be prepared to receive questions that you never thought anyone would ask of you. Don't react with exasperation or be rude. Always keep in mind that science journalists actually *do* care about science even though,

[2] The community's water supply was infected with bacteria originating from a nearby farm and several residents had died.

more often than not, mass media proceed to relentlessly dramatize the facts to promote sales and enhance ratings. The reporter who has a scientific assignment usually takes it very seriously, both from the journalistic point of view (reporting effectively) and the scientific point of view (being accurate).

Media Training

Media training is particularly helpful in preparing you for interviews. Most organizations have a communications department; some have in-house trainers; others hire trainers when needed. Check to see if such training is available to you.

Another way to prepare is to ask a colleague to put you through a mock interview where he or she deliberately asks questions to take the conversation down a different track, so you can practice bringing it back to the topic. Just as when you are preparing for presentations, it is very helpful to have your mock interview (if you can arrange one) or the actual interview videotaped. Seeing yourself in action will give you clues as to how you can behave next time so that you give the best impression—of your topic, your organization, and yourself.

SUMMARY

Communication, whether verbal, written, or through body language or imagery, is a universal process of expression that is important to all disciplines, particularly science. Although everyone is naturally capable of communicating, developing superior communication skills is essential for a successful career in science: for conveying scientific discoveries, informing and educating society, and advancing the world in which we live.

REFERENCE

Barzun, J. *Science: The Glorious Entertainment*. New York: Harper Collins, 1964.

Chapter

10

• • • • • • • • • •

TIME STRESS

Dealing with the Stress Associated with Time Pressure

• • • • • • • • • • • • • • • • • •

Peggy A. Pritchard, Editor

> *Time is the coin of your life. It is the only coin you have, and only you can determine how it will be spent. Be careful lest you let other people spend it for you.*
>
> Carl Sandburg, American poet

WOMEN: THE CONSUMMATE "MULTITASKERS"

Women are particularly good at getting things done. We have a well-developed ability to "multitask" that is a considerable asset when balancing the myriad demands of our professional and personal lives, especially during "pinch periods."[1] So why is it that we often feel that there isn't *enough* time in the day, as though we're constantly racing to get things done, that there's so much left to do? Why do we feel that we'd be able to accomplish it all, if we could just *manage time* properly?

[1] Those periods in life when the factors necessary for balanced productivity are least favorable. (Adapted from a phrase used in ecology. See Canadian Wildlife Service and Canadian Wildlife Federation, 2005.)

We're operating under the mistaken belief that it is possible to achieve all that we wish or need to do in a given period of time. But we can't. However much we would like time to be elastic, we cannot create more, nor change the rate at which it passes.

The pressure to complete activities within a particular period of time is not necessarily negative. Most of us work better when we have targets to meet. Deadlines energize us, inspire us to action, help us overcome inertia (and procrastination), and motivate us to accomplish the tasks at hand. How close the deadline has to be to have an energizing effect is partly a function of our temperament and influences the degree of stress that we experience. Some work better with last-minute deadlines, while others are more effective when their deadlines are further off and they can plan their work and pace themselves.

Working "Harder" Doesn't Work

But working efficiently and managing multiple demands effectively cannot compensate for simply having too much to do. As the pressure to accomplish more than is possible in a fixed time frame intensifies, our anxiety increases and we experience "time stress." The consequences are very real. If we do not set appropriate boundaries (for ourselves and others—see below), and in our multitasking we expect to maintain an increasingly unreasonable pace, we will suffer burnout.

Time stress is not only of our doing. Systems that value and reward workaholism and evaluate productivity using predominantly quantitative measures (e.g., long publications lists, large research groups, absolute levels of funding) greatly exacerbate the problem. Estimates of how long tasks or projects will (or, worse, "should") take to complete can be woefully inaccurate. They often do not take into account the unpredictable, such as delays due to failures of equipment and their subsequent repair (e.g., a machine breaks down and the part has to be shipped from another country; our computer becomes infected by a virus that destroys the hard drive), interruptions (e.g., due to more important projects intervening, illness), and the like.

Since we cannot "manage" time, have only a modest influence over the expectations placed on us by our employers and others, and cannot anticipate the unexpected, we need to manage ourselves differently. We need to be more reasonable about our expectations and the commitments that we assume, more cognizant of the time and energy that really is available to us, and learn to negotiate greater reasonableness from others. Though we cannot control all the external demands placed on us, we can choose how we will address the pressures and control how we will respond.

Do the "Right" Things and Do Them "Right"

If I spend time on this activity, I can't spend it on something else.

To gain a sense of satisfaction in our lives, we need to spend our time—as far as possible—on activities that will help us to achieve our goals (i.e., do the "right" things). We need to develop awareness (e.g., of our priorities, goals, responsibilities, how we currently choose to use our time), then organize ourselves and our activities to do first things first, set appropriate boundaries (e.g., by choosing what we will and will not do, performing activities at an appropriate level), and relieve the associated tension and mental fatigue (i.e., do things "right").

A magic "system" for managing the stress associated with time pressure does not exist. Neither is it an innate ability. Rather, it is a skill that can be learned, that evolves with our changing commitments and responsibilities. Some scientists create lists of things that they need to accomplish in a week and then check off the items as they complete them. Others prefer to divide their days and weeks into time slots and schedule key activities into specific slots. Still others work better with a more relaxed, fluid approach, keeping in mind their important commitments and fitting in other activities when they can. What works well for one person will not work well for everyone. The point is to develop an approach that is effective for you and to modify it as your situation and circumstances change. The following strategies may give you some ideas and inspire you to develop an approach that works best for you.

DEVELOP AWARENESS

ΓΝΩΘΙ ΣΕΑΥΤΟΝ (know thyself)
Inscription on the Temple of Apollo at Delphi, Ancient Greece

Life is "messy." No matter how we may try to organize, plan, and control things, there always will be interruptions, crises, urgencies, divergent interests that interfere with what we're doing at the moment. These diversions can hinder and even prevent us from doing what we must do and what we want to do. This is normal, just as it is quite natural to become so caught up in the daily demands of work and life that we begin to believe that everything must be given equal attention, immediately. But of course, not everything is important, however urgent it "feels."

To effectively manage time stress, it is helpful to begin with an awareness and understanding of your personal style and how it relates to time stress; current use of time, how long activities take, and the kinds of things (internal and external) that distract you from your tasks; and (most importantly) your professional and personal priorities.

What Is My Preferred Way of Working?

We all have preferred ways of doing things. We differ in our energy levels, biorhythms, and tolerance for and ways of coping with stress. When we understand what works best for us, we can make better choices about how we will use our time. Some questions to ask yourself about how you prefer to work include:

Do I prefer to work on one task at a time, completing it before moving on to the next? Or do I like to have several things on the go and move quickly from one to the next and so on?

Am I energized by last-minute deadlines? Or do I prefer to complete tasks well ahead of time?

Do I work in intense bursts of effort? Or do I prefer to work evenly and regularly?

When faced with work that has easier and more difficult parts, do I tackle the difficult aspects first to get them over with so that I can enjoy the easy parts? Or do I like to warm up, build my confidence and momentum with the easier aspects, and then move on to the more difficult parts?

Another aspect we need to understand is how our energy levels change throughout the day (week, month), when we are most alert and productive, and when we need to take a break (i.e., our biorhythms).

What time of day am I most/least effective? When do I do my best work?

How often do I need to take a break?

What activities help me to relax and recharge my batteries?

What is reasonable for me to expect of myself in a typical day? Week?

The answers to these questions and others will provide important information that will help you to make the best choices about what and how many commitments you can comfortably assume and how to organize yourself and plan your activities so that you will work in ways that are most consistent with who you are and how you work best. For example, many scientists arrive at work earlier than their colleagues (and students, if they're in academic environments) so that they have uninterrupted time to read the scientific literature or work on manuscripts. Others work from their home offices one day a week (see the section "Establish your absence"). Some keep strict office hours and never answer incoming calls during meetings. Others make an appointment with themselves every lunch hour to exercise at the gym, pool, or arena.

How Am I Using My Time Now?

Sometimes we have difficulty managing our myriad responsibilities because we do not have a realistic idea of how long it will take us to do things. This can be solved. What better way to gain a realistic perspective on how our time is spent than by keeping track? By observing and recording our current behavior during a "typical" period, we will be able to collect objective "data" on what we do, when, and for how long, that will help us to understand the multiple demands on our time and the frequency and nature of interruptions, and identify what strategies are working and what can be modified or eliminated entirely.

Keep a "Daily Activities Log"

Collecting data for this kind of analysis is simple and need not require extra time and effort. In fact, the time and energy that you save by making adjustments to how you do things as a result of your analysis is well worth the little time you spend in collecting the data. There are many possible ways to chart your work (see below for an example), and you will develop an approach that works best for you. The important point is that you record enough detail for a sufficient length of time (minimum of one week) for meaningful analysis. You may wish to keep a log during a typical period *and* during pinch periods, so that you have a record of your choices under different circumstances and degrees of time stress.

In the sample log that follows (Table 10.1), the day is divided into half-hour time slots (beginning at 8:30 a.m.) with columns for recording your activities, comments, and total time on task (t). Some people keep track of their activities for the entire day (to analyze personal and professional activities); others prefer to limit it to their working hours. Choose a time interval that is most appropriate to your circumstances. Obviously, shorter intervals will result in more data from which to draw conclusions, but this exercise is not intended to be burdensome. Set up something that is reasonable and that you're confident you'll complete.

As you proceed through your day, record what you've done. You may wish to record in the "comments" column what you thought or how you felt

Table 10.1   Daily Activities Log

Date:

Time	Activity	Comments	t
8			
8:30			
9			
9:30			
10			
10:30			
11			
11:30			
12			
12:30			
1			
1:30			
2			
2:30			
3			
3:30			
4			
4:30			
5			

about what you were doing since this can greatly influence your stress levels (as you'll read in the chapter "Mental Toughness").

As you analyze the log data, look for patterns in your behavior and that of others. For example,

Are there any patterns in how I spend my time?

Am I more productive at some times than others?

Are there more interruptions during certain times?

How am I avoiding doing uncomfortable or difficult tasks through busywork?

Is the time I'm spending on activities reasonable?

Under what circumstances did I feel best about my use of time?

What activities helped me to decompress or renew my energy level?

Once you have a more realistic idea of how you're currently using your time and have used the data to reinforce or change some of your choices and behaviors, you need not continue keeping a log. If your situation changes and you need more information about your patterns and choices, you can always return to this exercise.

However, you may wish to continue keeping a log—perhaps in some modified form—as a tool for planning your day and staying on track. It need not be a tool for evaluation and improvement only. It also can be a valuable source of objective proof that you actually *are* making progress. The daily work of science can be very tedious, with few immediately tangible results. If you're the kind of person who needs a sense of accomplishment each day or who tends to become so focused on what is still left to do that you don't remember what you have completed, keeping track of important measurable accomplishments on a weekly or monthly basis may be one way of reassuring yourself that you are making progress. For example, if you continue to keep a log, you'll be able to record the number of hours that you dedicated to planning experiments and writing manuscripts, total pages written, number of students tutored, etc. Collecting this information on a regular basis will also simplify the preparation of monthly or annual activities reports.

What Are My Priorities?

When you are clear about your priorities, your choices become obvious and decisions clear. Decisiveness equals confidence.
    Susan Wood, Associate Research Director, BIOCAP Canada Foundation

No matter how well organized, efficient, and capable we are, we cannot do everything that we need to do, let alone what we may wish to do. If we do not have a clear idea of our priorities and how each potential activity relates to them, we may well spend most of our time solving crises, meeting other people's needs and agendas (rather than our own), or becoming mired in busywork. In science, we may ask ourselves "From among all the possible questions I could ask about my subject, on which will I focus?" Likewise, in priority setting, we may ask "Of all the possible activities, commitments, and responsibilities that I could become involved in, which ones are consistent with my values?" It is our priorities that determine what is important and what is not.

In his widely acclaimed book, *The 7 Habits of Highly Effective People,* Steve R. Covey (1989) identifies *two* key factors in setting priorities: *importance* and *urgency*. By importance, he means how closely the activities relate to our professional and personal priorities, that is, their significance, value, consequence to our lives. The second factor—degree of urgency—refers to how pressing is the demand for action or attention (i.e., the closer the deadline for completion, the greater the urgency).

Obviously, we can rank our activities according to each criterion separately, but by examining them from both perspectives and charting them on a 2 × 2 matrix (Table 10.2), we will be able to identify more clearly and graphically those that deserve our attention and those that are best avoided. The names of the four quadrants make this obvious: "Important and Urgent," "Important and Not Urgent," "Not Important and Urgent," and "Not Important and Not Urgent." It is items in the latter two quadrants that are time wasters and busywork.

Covey's work was originally developed for the business community, but it is easily applicable to the responsibilities inherent in a life of science. Dr. Kathy Barker, in her excellent laboratory handbook for early career scientists (2002), demonstrates the general categories in a scientist's life and the quadrants in which they belong (Table 10.2).

You can use this approach to clarify your own priorities. Depending on the complexity of your life, you may have to rank the items within each quadrant. If you're having difficulty doing so, you may wish to apply a process that Patricia Hutchings (2002) terms "paired comparison." Consider each item, one at a time, with all the other items in the group and ask yourself this question: "If I had time to do only one of these, which would it be?" Put a check mark beside that item. Continue down the list, comparing each pair of items until all have been compared with each of the others. Count the number of check marks that each item received, then put them in order by score (highest to lowest). Your ranking is done.

Table 10.2   Time Management Matrix

	Urgent	**Not urgent**
Important	I Crises, personal or professional Pressing personal or equipment problems Deadline-driven projects	II Reading journals Relationship building Laboratory meetings Thinking and planning Recreation and relaxation
Not important	III Interruptions, some calls Some mail, some reports Some meetings Many administrative tasks	IV Trivia, busywork Some mail Some phone calls Most email and Web-surfing

As adapted in *At the Helm* (Barker, 2002). Adapted and reprinted, with permission, from Covey (1989)

Streamline and Simplify

What better way to decrease the stress associated with time pressure than to eliminate some of the responsibilities, activities (e.g., those that are unimportant, whether or not they're urgent) and relationships (especially negative ones) that are creating the pressure in the first place? By streamlining and simplifying our lives, we'll have fewer things to do, less to take care of and maintain, and fewer associations with negative people.

Getting rid of clutter will also simplify our workflow and make it easier for us to find things when we need them. By developing a workable, easy-to-maintain system of organizing and keeping track of information and items, we can reduce the frustration and embarrassment associated with missing important meetings, appointments, or other commitments and the time required to find misplaced items. There are myriad "systems" available to address almost every need: calendars, personal data assistants, scheduling software, filing cabinets, databases, reference management systems. The important point is to choose something that is consistent with your preferred way of working and is easy to maintain.

ORGANIZE TO DO FIRST THINGS FIRST

Now that you've identified your priorities, you'll be able to organize your activities so as to do first things first. This is particularly important during pinch periods, when you may have time only for these. The following strategies are just some of the ways in which you can accomplish this.

Set Goals

It is not enough to take steps which may some day lead to a goal; each step must be itself a goal and a step likewise.
        Goethe, German poet, dramatist, novelist, and scientist

Goal setting is one of the most effective ways of focusing our energies and making the best use of our time. The process requires a clear sense of our priorities (which we've already identified) and an ability to distinguish which activities are most likely to fulfill them. As scientists, we are accustomed to breaking problems into manageable pieces and tackling the components one at a time. So we already have the skills to set goals effectively.

The process can be as simple as creating a "to do" list for a set period of time or as elaborate as developing a comprehensive series of formally written goals and associated objectives.

"To Do" lists act as tangible reminders of what we want/need to accomplish, are simple to create, easily referred to, and provide tangible evidence of "success" when the items are checked off after completion. Scientists who prefer this approach often find that creating a list for a *week* is more realistic, brings better results, and provides greater satisfaction (because more is achieved) than creating daily "to do" lists.

For those who prefer greater structure, a more formal approach may suit. Hutchings (2002), for example, describes goals as having six characteristics: (1) written, (2) stated in the present (rather than future) tense, (3) phrased using positive language, (4) assigned a deadline for completion, (5) measurable, and (6) realistic.

The act of *writing* our goals helps us to clarify our thinking and makes our goals more tangible. Stating them in the present tense gives them greater motivating power because the language itself affirms that the goal is already achieved. Further, positive language helps us to focus our mental and physical energies on what *can be done* (i.e., positive action). It also helps us to avoid asking ourselves "unhelpful" or "dangerous" questions (see the chapter "Mental Toughness" for definitions) about our work and time constraints.

The importance of measurable and realistic goals cannot be overemphasized. Obviously, if a goal can't be measured in some way, we won't be able to judge effectively whether it has been achieved. And creating unrealistic goals will only serve to discourage and demoralize us rather than motivate and encourage us.

As mentioned earlier, assigning a deadline is another obvious way to motivate action, for it specifies a time when we will be called accountable (even if only to ourselves). A stronger motivator would be to enlist the support of a colleague or friend by sharing our goals with her and asking her to follow-up on our progress toward our stated deadlines.

The next step in the more formal planning process is to break down each goal into smaller steps, or objectives, that have the same six characteristics just described. With our goals and objectives thus established, we are in a better position to plan how we'll use our time and organize our days in order to achieve them.

A note about judging performance. If you are a person who tends to judge herself by her performance only, use criteria defined by the priorities and goals that *you* used when setting the goals in the first place. It was these that guided your choices and actions. If you judge yourself by any other criteria (especially criteria established by others whose values and priorities differ from your own), you are bound to fall short of success. This could lead to feelings of inadequacy that may undermine your confidence and self-esteem.

Plan How You Will Use Your Time

When estimating how long it will take me to do something, I like to use the following rule of thumb: I estimate the time as well as I can, then double it.

As suggested earlier, as far as possible, work in ways that best suit your personal preferences, style, etc., and that are most sensitive to your professional and personal commitments and responsibilities. If you have a plan in mind or, better yet, on paper, you're more likely to accomplish what you set for yourself. A plan need not be something rigid and unchanging—this would serve only to create more stress—but something to be used as a guide, as a "default mode setting."

One way to plan how you will use your time is to create a reasonable schedule (on a weekly, monthly, yearly basis—whatever suits you best). The schedule can be as simple as committing yourself to work on activity x in the mornings and y in the afternoons, with mail and phone message checks at the beginning of the day, at lunch, and before you leave. Or it can be more elaborate, with your day (week or month) broken into specific time slots to which activities are assigned. Remember to schedule time for important but not urgent tasks as well as those that are important and urgent. Minimize the time that you allow for unimportant tasks (however urgent they seem) or eliminate them altogether. (The chapter "Mental Toughness" may help you to learn how to "let go" of any negative feelings associated with these decisions.)

Unforeseen demands will interject themselves and demand your time and attention. This is a normal, albeit potentially frustrating or annoying occurrence, and you'll have to make adjustments as appropriate. You may even wish to build flexibility into your plan to accommodate the unexpected (e.g., the occasional "free time" slot). But at least you'll have something planned, so you won't waste time deciding what to do. And you can relax because you'll know you've set aside specific periods to accomplish the tasks that will further your own priorities.

You may wish to use a modified version of the "Daily Activities Log" (Table 10.1), create something new, or purchase one of the many commercially available paper or electronic organizing systems. Base your choice on what will suit your own needs, preferences, the demands of your work, and what you'll actually use. Your employer may require you to use an electronic system that enables all employees to access the schedules of others, so that group meetings, and so on, can be booked (e.g., Microsoft's "Corporate Time" program). If so, use it carefully. Again, make sure you assign time in your schedule for those important activities that are not urgent (e.g., reading, thinking, planning, writing). If you do not, your schedule will appear "open" and someone else will fill it. Be careful how you describe the commitment—to discourage others from assuming that their priority is greater than yours. For example, if you wish to block off time to plan a series of difficult experiments, do not write "planning time." Instead, enter something that can be construed as involving other people, for example, "research meeting." Granted, the appointment is with *yourself*, but your colleagues may think you're in a meeting with someone else and may be less inclined to ask you to reschedule it (or interrupt you if they want to see you immediately).

Organize Your Personal Life

It's difficult to maintain well-ordered professional lives when our personal lives are in chaos. Certainly, when we are balancing professional and personal spheres, there will be conflicting demands caused by overlapping activities and responsibilities. There will be times when the needs of people who rely on us (be they children, aging parents, or members of our extended family) will require our immediate attention. Interruptions and disturbances will be more frequent and numerous. (This is particularly true for academics, consultants, and other scientists who may spend more time working from home than do other professionals.) But we can give some order and structure to this aspect of our lives that will minimize the impact on our professional lives.

Just as you set professional priorities and plan your time, you can establish personal and family priorities and plan accordingly. For example, create a detailed calendar, agenda, or large chart that is posted on the wall or refrigerator to keep track of the activities and responsibilities of all the members of the family. Use multicolored pens to identify the person, activity, degree of urgency, etc. Involve everyone in the process and in the upkeep of the schedule. Expect your partner, children, and/or other household members to contribute to the maintenance of the home. It is particularly important to set appropriate boundaries (see the section "Minimize Interruptions").

Create Agendas for Discussions and Meetings

We all know how much time can pass in casual conversation around the coffee machine and in unfocused meetings that, in retrospect, were a waste of time because they did not contribute to the fulfillment of our goals. Networking and collaboration are two of the marvelous aspects of science that can enhance our performance and enjoyment of our work (the chapter "Networking" discusses other many benefits), but when time pressure is an issue, it is important to know how to set boundaries for our conversations and manage our interactions so we stay focused on achieving our goals. One technique is to have an agenda (however informal or unspoken) for each interaction. In meetings, a well thought out agenda will define the purpose, address salient issues, eliminate irrelevant topics, and help you and your colleagues to stay on track (as long as the chair of the meeting facilitates it properly). Even in informal discussions—for example, during a coffee break at a conference—it is possible to direct conversations (e.g., through careful questioning) to focus on topics of greatest interest.

This technique can work in one's personal life as well. A written agenda for family meetings, for example, will give children a feeling of being "grown up" and help them focus on the topic at hand (there's a paper in front of them). With a greater sense of membership comes a sense of responsibility as they learn that "their" items are important enough to be on the agenda. When they are younger, parents will facilitate/mediate the meetings, but as they grow into adolescence, children can take on more responsibility, learning to "chair" meetings. Not only can this make family discussions more focused and productive, it will be more enjoyable for all.

This is not to suggest that every single human encounter should be scheduled, regulated, or orchestrated. On the contrary, there is tremendous value in sharing coffee and conversation, meeting quickly for informal chats, having unstructured family time, etc. The point is this: Be aware of how you are using your time when you are interacting with people, professionally and personally, so you can make the best choices for the circumstances.

SET APPROPRIATE BOUNDARIES

A useful strategy for decreasing unimportant activities and responsibilities and protecting ourselves from distractions, interruptions, and the demands imposed on us by others is to set appropriate boundaries and stick to them. Some people seem to be able to assert themselves with ease and are comfortable letting others know that their time is limited and precious. For others, it can be more difficult. People who tend to be nurturing may have difficulty putting their own needs (professional and personal) ahead of those of others. And those who have low self-esteem may tend not to value their time or respect their own priorities. But it is a skill that can be learned, something that, in the beginning, mentors and members of our network can help us to develop. With experience and practice, we will gain confidence in our judgment and in our ability to discriminate, quickly and accurately, between demands that will contribute to the fulfillment of our goals and those that will not. And by helping those who work for us (and live with us) to do the same (for example, by helping graduate students integrate into their schedules regular periods of reading the literature and writing or by limiting the number of extracurricular activities that children can choose), we will be teaching them valuable life skills and will decrease the pressure on our own time. The following are but a few strategies for establishing and maintaining boundaries.

Avoid Becoming a "Rescuer"

Your lack of planning does not constitute a crisis in my work plan.
Henry Adams, American historian

Though it is very satisfying to help others solve their problems and flattering to be considered an approachable advisor, beware of becoming a "rescuer" of people who do not take responsibility for their own lives. You need not stop what you're doing to immediately attend to the demands that others are placing on your time and attention. Instead, step back from reacting automatically (using the strategies for disengagement described in the chapter "Mental Toughness") and analyze the situation. Ask yourself, "Whose 'problem' is this?" "How will this activity advance my own agenda?" If their requests have little to do with *your* priorities and responsibilities, gently and diplomatically decline to become involved. Do not allow yourself to assume responsibility for fulfilling their priorities or for solving the problems that they may have created for themselves by their own lack of planning.

Certainly, providing support and assistance to colleagues in need is part of being collegial, is an important aspect of mentoring, and has its own rewards. However, if you recognize a clear pattern of rescuing others, you may wish to make some different choices. Your behavior may also be making the situation worse for your colleagues because, by becoming dependent on you, they have avoided learning the skills required for working productively and independently. And, you will find it more and more difficult to say "no" to the last-minute demands of others because a precedent has been set.

Say "No"

Saying "no" is easy. It's sticking to it that is the difficult part.
Colleen Cavanaugh, Professor of Biology, Harvard University

When you understand your values, priorities, and goals, knowing what you reasonably can and cannot manage becomes clear. When new opportunities (or demands) present themselves (or are thrust upon you—as is often the case), the decision to say "no" is much more obvious. But having decided against committing yourself, it is essential to stay firm, even if you have some doubts. Be consistent, so that others will learn that when you say "no" you cannot be coerced into changing your mind. In essence, you need to teach people (your students, colleagues, supervisors, managers, family) that when you say "no," you mean "no." Period. If you reverse your decision under pressure—even once—others will learn that it is possible to persuade you to change your mind. You don't want this extra pressure. And you haven't time to waste on such discussions. So at the outset, take the time that you need to make an appropriate decision. Ask for as much information as is required, pause to reflect, assess, and decide. And once you've announced your decision, stand by it.

But how do I deal with the increased pressure to do what they want? With the pressure to change my mind? One approach is to develop "stock phrases" to repeat, over and again, until the questions stop. These need to be short and succinct and leave no room for you to be manipulated. Do not feel compelled to explain your reasoning. Say as little as possible and continue to be polite and respectful. The more you explain, the weaker your position becomes. The requester may be able to counter your objections with (seemingly reasonable) solutions that will make it more difficult for you to avoid the commitment. And your rejoinders may give the impression that you are making excuses or offering rationalizations. Try to find some point of agreement first, then say "no" again. The aim is to wear down your questioner with answers that say "no" means "no."

For example, you have just been approached by your colleague from another department to participate in a committee that is peripheral to your interests and job responsibilities. After careful consideration, you have declined graciously, but it has not been accepted. You are now fending off the attack.

> You: *Thank you for the opportunity. I'm sorry I do not have the time.*
> Them: *This is a very important committee that would benefit from your expertise...* (flattery)
> You: *It is an honor to be asked. Unfortunately, I have other commitments that make it impossible for me to participate.*
> Them: *The committee needs input from diverse perspectives...* (appealing to your gender, race, disability, etc. rather than your expertise as a scientist)
> You: *I am sorry, but I cannot help you at this time.*

For this technique to be effective, you have to become comfortable with sounding like a "broken record" and with the potential looks of disapproval and "pregnant silences" of the asker.

Another strategy is to suggest that another person take your place. Or, if you are interested but the timing is inconvenient, suggest an alternative approach that would make it possible for you to participate. The aim is to deflect the focus away from you or change the conditions so that participation is more convenient for you. For example,

You: *Dr. Odori has considerable experience in these matters and would be a better candidate than I. Perhaps she is available.*

You: *My schedule is booked until the end of the March. I would be available to give the seminar after that.*

If it becomes patently clear that this is one responsibility that you cannot avoid, change your tactics and enlist the questioner's assistance in managing your increased responsibilities. (This approach can only be used with those who have power and authority to make decisions about your workload, such as a team leader, supervisor, or manager; or those who share the responsibilities, such as colleagues and spouses.) For example,

Them: *As head of the project, I want you on this work team.*

You: *I'm currently heading projects x and y and am involved in a, b, c, and d. What are your priorities for these activities so that I may know which to let go in order to allocate enough time for this new responsibility?*

Pause Before Committing Yourself

If you find it difficult to say "no" immediately (even when you know that it is the best answer), develop techniques for firmly (but politely) delaying the decision. No matter how much someone is pressing you for an immediate answer, they usually are willing to wait 24 hours. As with saying "no" immediately, have practiced phrases ready. For example,

I need to give it serious thought before deciding.

I don't have my calendar with me; I'll have to check it and get back to you.

I'm just on my way to a meeting (or appointment or lecture). Can I let you know tomorrow?

Once you have disengaged yourself, examine the invitation and evaluate the consequences of accepting and declining. Ask yourself how this extra involvement will contribute to the fulfillment of your professional and/or personal goals. Is this really something you want to spend your time and energy on? If your experience and instincts suggest that you should say "no," they're probably right. Don't waste any more time thinking about it. Take action immediately. It is much easier for you to decline and stay firm when you've just made the decision and are sure of yourself and your reasons.

But be strategic. As much as possible, let the person know your answer without actually speaking to them. If you wait for them to contact you again, you'll have to speak with them on their terms and you may succumb to their persuasions. They may, for example, catch you at an awkward or inconvenient time or when you're feeling distracted or vulnerable, and you'll be in the same situation as before: feeling pressured to do something that they want you to do and feeling uncomfortable about declining. So avoid speaking with them directly. This may sound cowardly, but people can be very persuasive, especially if they have more power than you. Strategies include sending your regrets by e-mail, placing a handwritten note in their mailbox, or leaving a voice mail message. To minimize the risk of catching them in their office, you can call early in the morning (if they rarely come to work before you), during the weekend, or when you know they're in the laboratory or attending a meeting. Some telephone systems enable a caller to connect directly to voice mail; do this and the phone will not even ring in the office.

If a face-to-face answer is unavoidable, use the strategies discussed above: Have a short answer prepared. Do not go into elaborate explanations or give excuses or allow yourself to be drawn into answering questions about your decision. Instead, continue repeating your short answers, gently, politely, firmly. Use them like a mantra. You'll be amazed by how this wears people down and encourages them to abandon their attempts to change your mind.

Delegate

What better way to relieve the pressure of having to complete a task yourself than to delegate it to someone else, even if they assume responsibility for just a portion of the work? If done appropriately, delegation is an excellent way to increase your productivity and effectiveness and decrease time stress.

It's important to know what to delegate, to whom, and when. Delegation is not about getting someone else to do something that you don't enjoy doing; it's about having them do something that is a better fit for their skills, knowledge, and experience. Think of it in business terms: how much are you worth per hour? Is this activity worth spending *your* time on (however enjoyable you may find it), or could you delegate it and turn your attention to activities more appropriate to your training and expertise? Could your postdoctoral fellow prepare a review of the literature or draft a conference poster, for example? Could your partner, spouse, or children take on some of the responsibilities for planning and making meals? Can you afford to hire a fourth-year student part time and a cleaning service for home?

You do not have to be in charge of a staff in order to delegate. Consider the members of your network. For example, for writing projects (e.g., manuscript preparation, grant proposals, conference talks), can you share some of the load by "delegating" to qualified colleagues the task of critiquing your drafts? By soliciting input early, your colleagues can help you to focus your writing more quickly and save you valuable time and frustration.

One of the concerns you may have about delegating your work is that you don't trust the person to do the job well. This is where proper choice, training, and trust come in. First, it is important to choose the appropriate person/people for the job. They must be capable and competent and have the time to do what you ask. Since this may be something that they don't normally do, they will need instruction and guidance. This is a training opportunity for them, so treat seriously your responsibilities as a mentor. And finally, you have to stay focused on the desired outcome and let go of any expectation that you may have that the result will be exactly what you would produce; this is not possible.

Minimize Interruptions

Though interruptions are not always unwelcome, too often they are. If you begin to understand when and how they occur and can identify patterns in the behavior of others—and your own—that increase the likelihood of their occurrence (by keeping a log, for example), you will already have substantial "data" from which to develop strategies for minimizing interruptions. There are many approaches that are reasonable, appropriate, and workable. The following are but a few that successful scientists have found to be effective.

Begin work early in the morning, before everyone else arrives, so that the corridors and telephones are quiet. One very productive scientist (who has always been an "early bird") is at her desk at 6 a.m. and spends the first two hours of every day writing. Or stay later than everyone else. Many scientists keep unusual hours (due to the nature of their research), and some organizations prohibit after-hour access, so this late-night strategy may not be an option. But for some, it may be a good solution. Shut your door and do not respond to knocks. Another approach is to put a note on your closed door requesting that you not be interrupted for a specified time period or that interruptions are allowed only for a crisis.

If you share office space with others, try to have your desk as far from the door as possible so that when someone enters the room, you are not the first person whom they see and want to interact with. Create privacy with a screen or divider, so you will be less distracted and others will be less inclined to notice and interrupt you.

Set regular office hours for visits by students, company representatives, and others. As much as possible, train them to make appointments. It's important to note here that having an unbounded, "open-door" policy can become problematic—for you and the people you are hoping to help. It's important to be available for consultation when there is a crisis, but it can become quite disturbing to have to deal with interruptions caused by someone who is overly dependent on you for support. Students and trainees, in particular, need to learn to organize their time. If you allow every interruption from them, you will be reinforcing behavior that will not be helpful in their future careers.

Use the same strategies when you work from home. Create a separate, private workspace, if possible. Teach the people with whom you live that you are not to be interrupted if you are at your desk or the office door is closed. If your children are young, you may wish to post a sign with a red light (do not disturb) or green light (I'm working, but you may interrupt if it's urgent). Install a separate telephone line and Internet connection, so you are not disturbed by telephone calls for other family members or competing with others for access to electronic resources.

Establish Your Absence

In a Web column of *The Chronicle of Higher Education* (1999), Richard Reis talks about the importance of "establishing your absence." He maintains that, as important as it is to one's working relationships to be present and to contribute, it is equally important for effective work management not to be available all the time. By being absent from your office regularly (e.g., by working at another location within the institution [such as a library] for an hour or two, or from home half a day a week), people become accustomed to you being productive, but not in your office or laboratory all the time. The advantages are twofold: First, you have regular blocks of uninterrupted time in which to do your work. Second, the people with whom you work learn that they cannot expect you to be there whenever they want you; they have to be more respectful of your time constraints and plan ahead when dealing with you. When practiced in moderation, with respect for your responsibilities and obligations, this can be a very effective strategy for decreasing the stress associated with time pressure and increasing your productivity. By valuing your own time in this way and honoring your own priorities and needs, you'll be more productive and derive greater satisfaction from your accomplishments.

Use Technology to Serve You

In theory, technology was developed to serve us. But too often, we allow ourselves to become carried away with the many possibilities that it presents. Be aware of how you are using technology and whether its attractions are wasting your time.

E-mail

Limit the number of times you check your e-mail each day. By leaving your mail program open, you are inviting interruptions and distractions that are avoidable. Use a spam-blocking software to minimize the junk mail that you receive. Examine the incoming messages: do they really need to be dealt with? If not, delete them. When creating a message, limit the length; if more detail is necessary, attach a file. (The chapter "Communicating Science" has other suggestions for e-mail composition and etiquette.)

Internet Browsing

When you're using the Internet in your work, be aware of your tendency to follow interesting but irrelevant links and limit it as much as possible. When you find sites that spark your curiosity, bookmark and save them in a separate folder (e.g., named "Curiosities") and return to your work. When you take a break, you can return to them and surf at your leisure. The point is to stay focused on the task at hand while you are working and play on your own time.

Delegate information retrieval as much as possible (e.g., to a graduate student or senior technician). However enjoyable, is it the best use of your time? Perhaps it is—your expertise may be needed to identify the best sites and the relevance of the information—but perhaps it is not.

Voice Mail

If you need a period without interruption, set your voice mail program to pick up messages without the telephone ringing in your office. Use other people's voice mail to return calls or answer queries (e.g., to decline an invitation to participate in a committee) and avoid long discussions. On the other hand, some people prefer to answer their telephones because they wish to deal with the questions immediately and avoid having to follow up. From your own observation of the kinds of requests that come by telephone, decide which approach is best for you. As with e-mail, check your voice mail only certain times of the day. You may wish to set aside a block of time to reply.

Properly Maintain the Technology

Interruptions are not always caused by people. Failures in technology (e.g., hard drive crash, broken centrifuge, dead car battery), insufficient supplies (e.g., ink cartridge dry, no more reagent, no clean blouse to wear to an important meeting), etc. all can inhibit productivity. So it is important to establish regular backup and maintenance procedures in your office and laboratory and at home.

Work Mindfully to Decrease Stress

In addition to setting boundaries to protect ourselves from the intrusions of others, we need to set boundaries for ourselves. It's normal, for example, to prefer spending time on activities that we enjoy and can do well than on those that stretch our abilities or challenge our self-confidence. But if we habitually invest ourselves in the former—as a way of feeling busy but avoiding uncomfortable tasks—we will undermine our productivity and effectiveness. Balance and discipline are required. We need to work "mindfully," as Dr. Robert Boice terms it in *Advice for New Faculty Members* (2000). The following are just some of the excellent strategies that he suggests.

One approach is to work on difficult tasks in brief, regular periods at times when we usually do our best work, then "reward" ourselves by switching to something more enjoyable. For example, you may commit yourself to spending an hour working on the next section of a grant application, after which you will reward yourself by reading the interesting article in the most recent issue of your favorite journal. You'll find that by working this way, you will be able to sustain your efforts and fulfill your multiple responsibilities, while experiencing less emotional tension—because you'll be making modest progress on all fronts. It's like pacing yourself when training for, and competing in, an endurance race.

Begin early, to minimize procrastination. So often, "beginning" is the toughest part. Writing a paragraph at a time, complemented by focused reading and more thinking about the work and the writing, will guarantee progress. And success. Stop sooner to moderate perfectionism. Even though you may feel as though you could go on forever, there will come a point of diminishing returns (and fatigue) that could undermine your efforts. (See the chapter "Mental Toughness" for further discussion of procrastination and perfectionism.)

Train yourself to be productive during the short periods of time that occur during the day. It is the rare job that routinely has long periods of uninterrupted time. Always carry a current professional journal, for example, to read while you're waiting for a meeting to begin or doctor's appointment, while riding public transportation, etc. Learn to quickly become engaged in what you're doing. For example, begin each writing session by rereading the final section you wrote in a previous sitting so that you can pick up the train of thought and move it forward.

As suggested in the section on developing awareness, time your activities to suit your preferences (as much as possible). If you're an early-morning person, for example, plan to tackle challenging and/or detailed work then and save the more mindless work for periods when you're not at your best. Pace yourself. For the "joggers," this means working on something at regular intervals over long periods of time. For the "sprinters," this means working intensely on one project for a longer period of time before changing focus.

"Disengage" as quickly as possible from time wasters. Sometimes we find ourselves in situations that we feel are wasting our valuable time and/or energy and that we find difficult to leave. There are techniques that we can learn. For example, during a telephone conversation with a long-winded person, listen quietly, without comment. Avoid giving him or her any verbal cue to continue (e.g., do not say "yes..." "um hmm..."). At the first opportunity to speak, ask the person to put the details in an e-mail message and say that you'll get back to him or her after you've reviewed the information.

If colleagues, business representatives, or students arrive at your door and it's clear they're inclined to chat, let them know that this is not a convenient time. Get up from your desk or move away from the bench to meet them at the door and talk there. Remain standing. Explain that you're busy and ask them to make an appointment. Suggest getting together for coffee or lunch. Or look at your watch, move into the hall (closing your door behind you), and appear as though you're leaving for a meeting. Whatever approach you use, be respectful, polite, and firm.

By setting boundaries–for yourself and others—you may be able to lessen the effect of guilt-driven participation in activities that don't relate to your priorities and to minimize the tendency to take on too much and hence prevent burnout. With a commitment to stay focused on fulfilling your own professional and personal priorities and goals, and patient practice, you will find a system that works best for you. The important thing is that you become aware of what's happening to inspire you to take on too much, change your approach and responses, develop strategies for supporting yourself through the changes, and then do things differently—one situation at a time.

DECREASE TENSION AND RELIEVE MENTAL FATIGUE

No approach to managing the stress associated with time pressure will be effective without specific strategies for decreasing emotional and physical tension and relieving mental fatigue. Though the topic is discussed in detail in the chapter "Mental Toughness," it bears revisiting briefly here. By developing mental "agility" and "balance" and incorporating into our *daily* lives those activities that will renew our physical, mental, and emotional resources, we'll become more effective at staying focused and pacing ourselves, and more able to quickly and effectively defuse tension during pinch periods. The benefits to our professional and personal lives will be enormous.

Develop Awareness

The first step is to recognize the signs and symptoms of mounting emotional and physical tension that are associated with time stress and the situations that engender these reactions, and to become aware of the thoughts that we have before, during, and after the "crises." These reactions of the "fight-or-flight" system of the human body, though normal, can inspire us to react mindlessly (rather than respond effectively). By identifying them early and stopping them from escalating "out of control" (i.e., so that we feel helpless in their "grip"), we will be able to "let go" and respond in ways that are more helpful and productive. And, if implemented successfully, the likelihood of us developing mental fatigue is greatly diminished.

Becoming aware of how much effort that we are expending and being able to assess the quality of the results are two more important awareness skills. We can save ourselves valuable time and energy by avoiding perfectionism and investing only as much time as is needed to produce the necessary results.

Let Go

Most of us have heard the saying by the Earl of Chesterfield: "Anything that is worth doing at all, is worth doing well." A corollary, suggested by Dr. Thomas B. Newman (professor of epidemiology and biostatistics,

University of California, San Francisco), helps me deal with time stress imposed by the priorities of others: "If it's not worth doing [but it has to be done], it is not worth doing well." So I complete the task to the best of my ability in the given time, let it go, and move on.

. . .of "Shoulds"

When we're under the pressure of deadlines, we simply do not have the time to dwell on what "should" be (e.g., "My manager should have asked me to do this long ago." "I should do everything well, even if it is not important, but I have to do it."), what "should have been" (e.g., "My colleagues should have finished their sections of this report by now." "I should have anticipated this."), etc. Though it is appropriate to analyze situations—so we can learn from them and improve our performance in the future—doing so *during* a pinch period is not the appropriate time.

. . .of Regret and Guilt

Regret and guilt are strong feelings that can cause us to react mindlessly and make it difficult for us to perform effectively when we're under the pressure of time constraints. They can be engendered by many situations in which we must make choices. Ironically, even circumstances over which we have no control may inspire us to feel a sense of regret or guilt, if our thinking is distorted. Letting go begins with accepting that these are common feelings to experience–that you're normal–and that becoming "immobilized" by them is unhelpful and can be detrimental to your self-confidence and self-esteem, your ability to perform optimally, and, ultimately, your success.

There will be occasions when we have to choose between two equally important (or attractive), but conflicting activities (e.g., meeting with the leading expert in your field who has unexpectedly dropped by your laboratory or giving your final lecture in a core undergraduate course). We need to make the best choice possible, accept that the other will not get done (or done immediately), and let go of the negative feelings. Techniques include (1) clarifying distortions in thinking (e.g., "I know I did not follow through on my commitment to my students, but this was an unexpected opportunity to help my research team finally solve this sticky problem. I'll arrange to deliver the information to my class in another way."), (2) engaging positive self-talk (e.g., "This was my final lecture and no one else could give it; I made the logical choice. I'll telephone Dr. X and speak with her at a more convenient time."), and (3) seeking the support of trusted members of our network (e.g., You express your sense of regret to a colleague who commiserates, "I understand the difficult choice you had to make...").

Another potential "trigger" of regret or guilt is making mistakes. Experiencing "failure" is a natural (and frequent) part of life—and the scientific endeavor. By choosing to accept and use mistakes as opportunities for learning, we're much more able to let go and move on.

. . .of Frustration and Anger

It is important to be able to release the emotional "energy" associated with strong feelings in ways that are appropriate and will not compromise your professionalism or personal integrity. This can be accomplished in any number of ways, depending on the situation and individual; it is helpful to have a number of options in your "stress-relief toolkit." An effective way to "let off stream" is to talk to a trusted mentor or friend about your frustrations. Some people enjoy a regular, end-of-week social time with a small group of colleagues in which they unwind from the stresses of the week. Others use humor (e.g., imagining the situation in a ridiculous extreme) to put the triggering situations in perspective and relieve the tension.

Vent Your Physical Energy

Many find that vigorous physical exercise is an excellent way to relieve the physical tension associated with stress and to clear one's mind of negative or obsessive thoughts that contribute to emotional fatigue. There are a great number of individual and group activities–recreational and competitive– that you can pursue, both indoors and out. Integrating them into your routine is most effective, for example, doing gentle stretching exercises or yoga in the morning just after rising, taking a brisk walk at noon with members of your laboratory, playing a game of squash with a colleague every Thursday, hiking with the community club once a month. The obvious added benefit of regular physical exercise is improved physical and mental health and increased energy. And if engaged in with some of your colleagues, important working relationships will be strengthened.

Establish Daily and/or Weekly "Decompression" Rituals

Long-term, single-minded focus and sustained concentration can lead to mental fatigue. Just as the body needs rest and relaxation, so too does the mind. By changing our focus periodically and establishing daily or weekly routines for helping us to "decompress" mentally, we'll be able to minimize fatigue. Examples include pushing away from our laboratory benches or computer screens every hour or so to stretch and look out the window, listening to a favorite radio station during the commute home, attending weekly choir practice, spending a quiet time each Saturday morning, reading the weekend newspaper, and relaxing with a glass of wine over a special meal with our spouse once a month. And, again, engaging in physical activity is an effective way to decompress.

Nourish Your Body to Support Your Mind

The importance of a balanced diet, regular physical exercise, and adequate sleep in dealing with physical tension and mental fatigue cannot be emphasized too much. Eating proper meals at regular intervals, in particular, can greatly influence your level of physical energy and mental acuity. Drinking enough water is also important, especially in buildings with sealed windows and dry, conditioned air.

CULTIVATE A POSITIVE ATTITUDE

Managing the stress associated with time pressure involves developing a realistic sense of what is possible to accomplish in a specific period of time, given our priorities, responsibilities, and constraints, arranging our life accordingly, and staying flexible enough to accommodate the unexpected. It's about saying "no" to things that are unimportant (however attractive), committing ourselves to doing first things first, and *choosing* to be at a peace with ourselves for the choices that we make, for our performance, and its outcome. When we make these strategies a part of our routine way of thinking, working, and approaching life, they'll become more natural and more automatic and will serve us well.

> *Everything can be taken from a man or a woman but one thing: the last of human freedoms—to choose one's attitude in any given set of circumstances, to choose one's way.*
>
> Viktor E. Frankl, psychiatrist and author

REFERENCES

Barker, K. *At the Helm. A Laboratory Navigator*. New York: Cold Spring Harbor Laboratory Press, 2002, p. 42.
Boice, R. *Advice for New Faculty Members*. Boston: Allyn and Bacon, 2000.

Covey, S. R. *The 7 Habits of Highly Effective People*. New York: Fireside, 1989, p. 151.

Canadian Wildlife Service and Canadian Wildlife Federation. *Hinterland Who's Who Glossary* (2005) [Online]. Available at: http://www.hww.ca/glossary.asp#letterP (accessed August 20, 2005).

Hutchings, P. J. *Managing Workplace Chaos. Solutions for Handling Information, Paper, Time and Stress*. New York: American Management Association, 2002, pp. 199–122.

Reis, R. "Establishing Your Absence," *Chronicle of Higher Education, Chronicle Careers Catalyst* (1999) [Online]. Available at: http://chronicle.com/jobs/v45/i42/4542ctlyst.htm (accessed August 20, 2005).

• •
SARAH E. RANDOLPH, Ph.D.
Professor, Department of Zoology
University of Oxford, ENGLAND
Parasite Ecologist
• •

Professor Sarah Randolph recently celebrated 31 years of unbroken employment on soft money, beginning with one short-term teaching post and followed by six successive personal awards of prestigious research fellowships. Her research support has come from the Leverhulme Trust, the Royal Society, the Wellcome Trust, and the Natural Environment Research Council, but not from the University of Oxford, where she has been based since 1974. Maintaining continuous funding under these circumstances is a significant accomplishment in itself; her contributions to the advancement of science, and to teaching and mentoring are equally so.

The old adage: "If I had not made these discoveries, then someone else would have done so soon afterwards," is *not* true of Professor Randolph's work. She has developed a uniquely detailed insight into the quantitative field ecology of pathogen vectors (first ticks and later tsetse flies), that has engendered entirely novel explanations for the epidemiology of the resultant diseases. These, in turn, offer informed predictions about changes in the risk of infection with changing environmental conditions. She is repeatedly invited to give lectures on the international stage, most recently as the Rausch Visiting Professor at the University of Saskatchewan, Canada.

Despite her status as a research scientist, Professor Randolph has always been totally committed to teaching within the university system, currently as tutorial fellow in charge of biological sciences at Christ Church (college), Oxford. She derives huge personal satisfaction from nurturing the development of the next generation and sharing with students the intense beauty and excitement of the natural biological world, analyzed within rigorous quantitative frameworks.

Professor Randolph would be the first to acknowledge the contributions made by her husband, Professor David Rogers (not least of which was to divert her from ticks to tsetse flies, so that they could do field work together in Africa). And if she were asked to reflect on her professional accomplishments, she would reply that these ephemeral performance indicators pale into insignificance beside her other life's work: Emily (1979), Thea (1982), and Jack (1984).

L. Cate Brinson, Ph.D.
Jerome B. Cohen Chair of Engineering
Northwestern University, UNITED STATES
Theoretical and Experimental Mechanician

Dr. Cate Brinson is an internationally recognized expert in mechanics of materials, working in such areas as mechanics of nanoparticulate reinforced polymers, characterization of titanium foams for bone implants, and modeling of shape memory alloys. Her investigations span the range from molecular interactions and crystallographic features to micromechanics and, ultimately, macroscale behavior. Dr. Brinson has received a number of awards, including the 2003 American Society of Mechanical Engineers Special Achievement Award for Young Investigators, a 2000 Alexander von Humboldt Research Award, and the Defense Science Study Group in 1998–1999. She has made numerous technical presentations on her research, authored over 50 journal publications and several book chapters, and is completing her first book *Polymer Engineering Science and Viscoelasticity.* She served on the Society of Engineering Science Board of Directors, including one year as President, and is currently serving on the National Materials Advisory Board of the National Academies.

An important role model and inspiration to those around her, she provides scientific mentoring and encouragement to her colleagues, postdoctoral fellows, and graduate students. In addition, Dr. Brinson is active in engineering curriculum development and undergraduate teaching. Her life is an inspiration to others, demonstrating that successful women do not need to sacrifice their personal lives to have successful, rewarding, and productive careers.

Dr. Brinson was greatly encouraged from an early age by her parents, an engineer and a mathematician, who claim she independently discovered Newton's third law when she was three years old. Musical interests and athletic talents complement her scientific passions. While oboe playing dropped by the wayside after graduate school, she still actively competes in triathlons and enjoys mentoring female students at the gym. Extracurricular activities also include three adorable children, house renovating, and many other hobbies. Her life's motto: "Yes, you can do this, and it will be fun!"

● ●
VALERIE ANN CORNISH, M.A., D.Phil.
Durham, ENGLAND
Biochemist
● ●

Dr. Valerie Cornish graduated from Brasenose College, University of Oxford (England) with a B.A. in Biochemistry, and then moved to the University's Department of Pharmacology, where she earned her D.Phil. investigating placental enzymic reactions and the metabolism of folic acid. A Wellcome Trust Prize Travelling Research Fellowship (1999–2002) enabled her to continue her study of early developmental biology, first at the University of Colorado (United States), where she focused on brain development in mice, then again in Oxford at the Department of Zoology. She subsequently continued her research in addition to beginning to develop her teaching skills back in the Pharmacology Department.

Two of her guiding inspirations have been women with a passion for science: her sixth form (grade 12 in the American system) chemistry teacher, and the supervisor of her fourth-year project (and subsequent D.Phil. supervisor) in the Pharmacology Department. She enjoyed the theoretical side of biochemistry as an undergraduate, but it was only when she began to put the theory into practice that she really thought it could turn out to be a rewarding career choice.

Dr. Cornish's interest in developmental biology is personal as well as professional. In October 2003, she took her research a stage further with the birth of her daughter, Charlotte. This coincided with a family move from Oxford to Durham, England, where her husband took up a new position. Since then she has been adapting to motherhood and settling the family into their new surroundings. When part-time work arose at the University of Durham, she investigated the opportunity with a view to establishing the right balance between professional and personal work and decided it was not the best choice for her and the family at the time. She expects to return to science at an appropriate time in the future and hopes it will be possible, despite her career break. Of her current research she says with conviction: "Charlotte is my best experiment yet!"

Linda S. Schadler, Ph.D.
Professor, Materials Science and Engineering Department
Rensselaer Polytechnic Institute, UNITED STATES
Materials Science Engineer

Dr. Linda Schadler is an experimentalist who has been active in research for over 15 years. A graduate of the materials science and engineering programs at Cornell University (B.S., 1985), and the University of Pennsylvania (Ph.D., 1990), she has known since she was in high school that she wanted to have an academic career. Dr. Schadler's research has focused on the micromechanical behavior of two-phase systems, primarily polymer composites, and more recently has broadened to include the mechanical and electrical behavior of nanofilled polymer composites. A member of the National Materials Advisory Board, her studies have resulted in more than 80 publications. Her contributions to science and education have been recognized through numerous honors including a National Science Foundation National Young Investigator award (1994); a Dow Outstanding New Faculty member award from the American Society of Engineering Education (1998); the American Society for Metals (ASM) International Bradley Staughton Award for Teaching (1997); the ASM International Geisler Award, Mohawk Chapter (2000); and a Rensselaer School of Engineering Faculty Research Award (2004).

Professor Schadler's commitment to education and mentoring is reflected in her involvement in outreach and education programs for high school students and teachers. She is the education and outreach coordinator for the National Science Foundation's "Center for Directed Assembly of Nanostructures," headquartered at Rensselaer Polytechnic Institute, and just completed a major project—the Molecularium. The Molecularium uses a dome theater (planetarium) to take children on animated magical musical adventures down into the world of atoms and molecules where they meet atom characters "Oxy" and "Carbone" and learn that "everything is made of atoms and molecules."

Linda was raised in a family of scientists, and the encouragement of her extended family has had a significant impact on her commitment to challenging the traditions of the scientific culture. Her husband and two children are a joy and continually remind her that "some things are more important than others," a phrase that guides many of her choices. As a family, they love to ski, hike, camp, and just spend time outdoors.

11

•••••••••

BALANCING PROFESSIONAL AND PERSONAL LIFE

••••••••••••••••

*Sarah E. Randolph, L. Cate Brinson, Valerie A. Cornish,
and Linda S. Schadler*

DO WE WANT TO LIVE TO WORK, OR WORK TO LIVE?

For sheer excitement when things are going right, there can be little to compare with science. Sooner or later, that first *Eureka!* moment will happen, and you will be hooked. You will cut through the haze of interwoven facts and competing ideas to an answer with mind-fizzing clarity, due perhaps to a new observation, a new experimental result, or detailed integration of disparate data and analysis. Whether the Eureka produces a small piece in the developing jigsaw puzzle or puts the last puzzle piece in place for a major breakthrough, it is this passion for discovery and innovation that drives the vast majority of excellent scientists on whom the health and wealth (in its broadest sense) of the world's peoples ultimately depend. As practitioners, we know that science is great.

Realistically, however, many of us feel that the scientists' way of life has become less than great in recent years. We are subject to increasingly intolerable pressures to do more, to produce more, to publish more (but only in certain high-ranking journals!). At the same time, increased productivity is impeded by additional increasing demands to multitask, not only as scientists and teachers for which we are trained and employed, but also as administrators, fund-raisers, accountants, and architects of building developments, at which we are reluctant amateurs.

It is important to recognize these increasing pressures because they can have negative effects not only on the balance between our professional and personal lives, perhaps turning younger scientists away, but also on our performance as scientists. Great science depends absolutely on great ideas. These develop best in an environment of spacious calm, with time to think and periods that may appear fallow but are in fact the wellspring of originality. The complaint "I can no longer think at my desk; I have all my best ideas as I walk home, dig the garden, cook supper..." has become a cliché only because it is so generally true. The importance of having a "life" outside "work" is incalculable, but for people who make their living by having ideas, the two are inseparable because, of course, we carry our brains with us wherever we go, whatever we do.

Not Just a Women's Issue

Our thesis is that all scientists are being caught up in a system that is increasingly making unreasonable demands on our personal lives. Traditionally, men have made sacrifices for the sake of their careers, missing out on leisure activities and much of their family life. Now that women are valued for their intellectual contributions to science, the conflict between personal and private lives risks excluding them from science simply because many women are not prepared to make similar personal sacrifices. Additionally, an increasing number of men are reluctant to be absentee fathers and husbands; those who are partners of women scientists are especially sensitized. Ideally, we would like to see radical changes in attitudes so that both men and women are encouraged to work shorter hours and have richer, more varied lives. This is not incompatible with a highly developed, affluent society. Sweden offers a good model in this respect, where more leisure time does not have negative economic impacts.

Meanwhile, we need to develop strategies to push the system and expectations in directions that promote balanced lives. Toward this end, we offer ideas on how to achieve balance between professional and personal life, especially when it involves family. Why do we emphasize family? Because it is this aspect of our personal lives in which fewest compromises are possible. By family, we mean any dependents for whom we have caring responsibilities. Parenthood is the extreme case, albeit the one arising most often from our own choice; it is unconditional commitment, 24 hours a day,

7 days a week, 52 weeks a year, for about 20 years per child. Its beginning also typically coincides with the time of establishing oneself as an independent scientist. That is why it often conflicts with or impedes a successful career in science and why it is important to work toward changing attitudes in the scientific professions.

At the same time, many of the general issues addressed here about achieving balance apply generally to men as well as women, to scientists without partner and children, as well as those with family. We all need to avoid getting ourselves trapped in a lifestyle and a time budget from which there appears to be no escape. The balance of work within our lives is something that we all need to be thinking about at all stages of our careers. We need to be consciously choosing our own personal work ethic at the beginning, and continually reassessing and adjusting through the different phases of our personal and professional lives.

> As the last of our three children effectively left home and the discipline of the school gate evaporated, I asked my husband whether we would ever leave work in time to enjoy the evening at home. I realized I was going to have to impose my own structure on the day if I was to avoid slipping into bad habits that I knew would be detrimental to my work and my home life. Ironically, just as I was faced with more time, I could foresee my life becoming imbalanced if I was not careful. S.E.R.

This chapter is written by four women scientists who have children and are in academia. Thus, the examples that we provide often reflect some of the detailed aspects of academic life. However, the life-balancing challenges facing women scientists are quite general, regardless of whether the work is in academia, government laboratories, industry, or other institutions. While it is impossible to write a chapter that contains the one and only solution, or even a menu of several sure-fire solutions, it is possible to outline clearly some of the defining issues for personal-professional balance for women in science and also to indicate some general strategies that have been successful. We hope in doing so that we can provide inspiration to others as they strive for their ideal balance, for each of us must come to our own decisions as to our center, our priorities, and our balance. In the following pages, we identify some important choices that have to be made about professional and personal directions and setting priorities. Throughout, we provide vignettes and practical advice and examples that can be starting points to challenging and changing the traditional system, success strategies to oppose the increasing professional pressures that give rise to imbalance. If we don't at least employ Newton's third law and meet those forces with an equal and opposite reaction, we will surely obey the second law and accelerate in the direction of imbalance!

FREEDOM OF CHOICE—DECIDING WHAT TO DO PROFESSIONALLY

"Which Would I Prefer: To Have to Choose Between Several Things I Want to Do or Have No Options?"

Freedom of choice is what balancing our professional and personal lives is all about. This "freedom" is by no means absolute, in terms of having unlimited options or being able to avoid any negative consequences of our choices; many decisions are forced on us by factors beyond our control (not including our gender and the associated cultural biases), and sometimes we may have to choose between alternatives that we consider to be "bad" and "worse." The point is to make choices *consciously*—with the best possible understanding of their consequences and in light of our current professional and personal values, goals, and interests.

Choosing to do what we feel will best suit our aspirations can be very positive and result in a very rich life; being able to express our inherent personal preferences and abilities is a privilege, a privilege not available to all. Choosing *first* to accept that we cannot do everything will help us to let go of unrealistic expectations (our own and others'), allow us to focus on our own professional and personal priorities, and enhance our confidence in the career and life decisions we do make.

> *It might sound trivial but when I was a child it came as quite a shock to me to realize that I was never going to read all the books ever printed. We all have to make choices about which books we read and which we don't. This insight was liberating! I then felt comfortable with my choices.* V.C.

Some of our "bigger" choices—over career directions and lifestyles, for example—have greater consequences but are not necessarily irrevocable. In today's global economy, where there are fewer jobs for life (and therefore less security) and a greater acceptance of a "portfolio career," it is easier to change directions.

> Science or the city? In Margaret Thatcher's materialist Britain, we complained that all our best graduates flocked to London with their first-class degrees, starving the research community of the next generation. The money was good and they no longer lived as poor students. But the hours were long and many quickly burned out. One of my recent graduates joined a leading investment management firm but grew to feel that his originality and analytical mind were not appreciated. He has taken steps to return to science and has the additional freedom that comes from a healthy bank balance. In the reverse direction, a very clever colleague found that the uncertainties and low pay of a postdoctoral fellowship were incompatible with

his new paternal status, and he chose to leave science. But he has subsequently returned! S.E.R.

But how does one decide? There are two highly personal steps to be taken. First recognize your own strengths and weaknesses, identify the benefits and drawbacks of the options that you're considering, and accept the compromises that must be made if you wish to achieve your (often competing) aspirations (e.g., between professional work and personal life). Second, choose to feel comfortable with the decisions that you make. This will derive from an inner confidence that what you are doing is right for you and also from being valued in whatever role you choose.

For me, identifying strengths and weaknesses comes relatively easily. Accepting compromise, however, is exceedingly difficult. When a conflict arises, such as an invited talk at a major conference versus my son's planned birthday party, I find myself trying to move both events so that I can do both. Last year I had to turn down an invitation to participate in a workshop run by the National Academy that conflicted with a complex vacation that had been planned for over a year. I still regret that I was unable to do both. The point is that while it is easy to say "set priorities and accept the compromises," it can be very difficult to carry that out. However, it is OK for this to be difficult. It is acceptable to have regrets. You are not alone, it is not debilitating, and you will continue to move forward anyway. L.C.B.

"What Would I Like to Do and Be Happiest Doing?"

What sort of scientist do you want to be? The purpose and pressures of the different career tracks in science will have implications for the balance between professional work and personal life. Within science, it is possible to choose a job with very specific work hours and goals, a job in which skills and training are directed specifically toward producing or developing a product with a specific use. Such jobs can be very liberating. For example, Lynne Richardson runs the gene-sequencing lab in our department, an efficient, reliable, and highly valued service. She purposefully switched track from research to this vital role when she had children, and says:

I organize my day and know that I am going to finish at the same time each day so that I can be home for teatime and homework. I set all the samples going first thing each morning, and then deal with all the paper work and accounts in the afternoon.

Most extreme in its potential to encroach on our personal lives is research science, within the commercial sector, a research institute, or a university. This is all about discovery, and there is no end to it, ever. In non-commercial establishments, salaries may come from the public purse, but still we work primarily for ourselves, for our own satisfaction. We are very lucky indeed to be paid for being allowed to enjoy ourselves so much! The passion and the drive to discover the truth are so pleasurable that it's more like playing a game against ignorance. Eventually, however, this can become competition between scientists to win the prize rather than to beat ignorance. The way that we cope with these daily driving forces depends very much on our personalities. Ask yourself, "How ambitious am I? And for what?" As scientists, we know within ourselves when we have made a significant contribution and whether our work is likely to have an impact on the lives of others (including our students). When you judge your impact to be positive on both counts, is that enough?

I can honestly say yes, for me this is enough. Recognition by one's peers is also nice, but this can hardly be called fame. Despite acknowledging my healthy competitive streak in sport and games, what I can't stand is the attitude that science is a zero-sum game, where winning by one scientist implies losing by another. Unfortunately, where resources—grants, faculty positions—are in short supply, this is exactly how it is. S.E.R.

Certainly, many scientists are driven by the lure of fame, power, and the fortune that may come with it. But there are very few of these kinds of "rewards." In some fields, or at certain, highly focused points in a line of research, a particular discovery is so crucial that the "winner takes all." "Runners up" go down in history as also-rans, if at all, despite the fact that their work may be as good or better. It is just that their discoveries were made and published at a different rate. How well will you cope with this pressure? Competitiveness and scientific talent do not always go together— there is no guarantee that you will "win." And the more competitive you are, the more vulnerable you will be to succumbing to the pressures because the outcome ("winning") matters so much to you. It is in this state that professional work may take over your life, to the exclusion of all else.

Some fields of research are more prone than others to frantic races against time. Cell and molecular biology is a prime example, because the methods, conditions, and outcomes are so exactly reproducible, and experiments can be planned, set up, executed, written up, and submitted for publication within days or weeks. *Nature* published an article featuring a duo that did indeed publish first (Pearson, 2002), but only by virtue of the husband moving into his wife's laboratory and their working together far into the night for 3 months. The article went on to bemoan the loss of freedom of exchange of ideas resulting from this increasing culture of competition and accompanying secrecy, not to mention the potential for sloppy science.

Is this the way you want to lead your life? From their autobiographies, it is very clear that Jim Watson thrived on the cut and thrust of the competition to crack the physical basis of the genetic code (DNA's double helix), whereas Maurice Wilkins—even before he found himself eclipsed by Watson and Crick—hated any idea of a race. You may prefer to select one of the many less frantic branches of science. Field biology, for example, is very different: collecting the data may take years, and the natural situation is so complex that no two sets of results are so identical as to render parallel observations or experimental results redundant.

But even outside the realm of aspiring Nobel Prize winners, the increasing pressures of a scientific research career can still be extreme. As mentioned earlier, we all feel enormous pressure to publish X papers per year, to compete for and obtain significant funding, to serve on prestigious committees, to contribute to research societies (from running sessions at conferences to being on the board of directors), to be the chair of the department; the list is never ending. Choosing where along the spectrum of achievements you will feel fulfilled is critical to finding the balance between personal and professional life.

> *In academia, I find that I am constantly presented with a myriad of opportunities: run graduate admissions, serve on promotion committees, be on the board of XX society, write YY proposal with a collaborator, be in charge of development of a new curriculum change.... And individually all these options are fun and desirable. It is the cumulative effect that is so draining. Though I love my research and work with my graduate students and postdocs, I harbor no illusions or aspirations toward major research recognition. Nevertheless, the pressures of my professional life do impinge on my personal life and I must exert a constant conscious control to maintain equilibrium.* L.C.B.

Life in Partnership: Choose Wisely!

> *I have been very lucky in having the right husband. Time and time again women who are successful in science have said this to me—you need both a passion for your subject and the right husband.*
> Dame Julia Higgins, quoted in *The Times Higher Educational Supplement* (Sanders, 2001)

Most of us seek to fulfill many of our personal needs through relationships. And it is within partnerships that we commonly, but by no means necessarily, find that extra dimension that gives life meaning and full enjoyment. Partnerships, however, may be the ultimate "catch-22." A life in science (or any profession, for that matter) is not a 9-to-5 job, but a vocation. And for some, it is a calling. Your partner needs to understand the importance of the work to you and the commitment of time and energy that you are making. There's no magic formula for working out the details; they have to be negotiated.

> *When I am asked advice, as I often am, by young women wanting to make a career in science, one of the things I tell them, half-laughingly, is to avoid falling in love and committing yourself to anyone until after you have established yourself. Get the job first, and then the partner. This is hardly useful advice because, of course, such things cannot be ordained, but it is the rueful recognition that emotional forces can complicate matters.* S.E.R.

It is difficult enough for one professional to secure a job in an organization, institution, or community, let alone two. Matching vacant posts in science to a specialist's skills and interests usually requires that one be free to move. When a choice has to be made between a partner and a job that best suits their training and abilities, it is more often women who put the priorities of personal life before professional goals.

> Q: "Valerie Cornish: If you could do it over again, would you make the same choice?"
>
> In terms of work-life balance, the first real priority decision that I had to make was on getting married. My husband and I had just completed our D.Phil. degrees, and he had accepted a postdoc position in Colorado. I had a six-month agreement to carry out research at the same university–at a "basic rate" of pay–with a view to my applying for funding to continue. Our priority was each other (a joint decision): it was more important to us to be in the same place than for me to hold out for the perfect job before we moved. My husband would be earning enough to support us, and the experience of living abroad seemed more beneficial than having two incomes. Fortunately, within three months of arriving, I had secured a three-year research fellowship. Without hesitation, I would do the same again, even if I had never obtained funding; our time in Colorado cannot be measured purely in terms of career advancement or financial gain.
>
> I am a great believer in asking for what you want. You won't necessarily get your own way, but if you don't ask, you certainly won't. When we first arrived in Colorado, I negotiated to work part-time (until my fellowship began), so that I could organize our new home (buying furniture, household goods, etc., without a car is no easy task!). I was also able to spend 10 days with my sister (living in San Diego); another luxury money can't buy. Admittedly, I felt I had some bargaining power due to my salary, but I doubt that the part-time position would have been offered had I not asked. When we returned to the U.K., I negotiated

a month's delay to the start of the third year of my fellowship, again because I felt that the time it gave me was more precious than anything else. Being at home to receive our furniture from the U.S., organize telephone connections, etc., smoothed our transition considerably.

I have now opted for a complete change in direction since the birth of our daughter in October 2003. This event coincided with my husband taking his research fellowship from Oxford to Durham; consequently, I resigned from my position. Our move was not an easy decision to make, especially once we discovered I was pregnant. I was in a reasonably strong position to negotiate a part-time return to work in my department, which is very supportive of this. However, we decided that the move was definitely best for us as a family.

Currently, I am relishing being a full-time mother and settling into our new surroundings. I did investigate a part-time position at the university here, but we decided it was not the right post for our family at this time. My Mum, who taught maths, resisted pressure to take on additional responsibilities at work when we were children because she knew that she wanted to be able to leave school when the bell rang and be at home for us. She saw people much less competent than her progress further up the career ladder, but she was rewarded in other ways. I too can accept that with similar priorities, I have to accept the consequences for my career. V.A.C.

It is interesting to note that women in science, engineering, and technology (SET) commonly end up in partnerships with other scientists, although the converse seems to be less common (see census statistics by Blackwell and Lynch 2002). The scientist-scientist partnership can be a winning combination, as long as the professional stresses, when shared, are halved rather than doubled. It can bolster one's confidence enormously and make the difference between success and failure. An added benefit is 24/7 access to an informal, cost-free, research "consultant" with whom to share and develop ideas.

I have benefited enormously from having a husband in science—he understands the career pressures and critical times, so I never feel guilty. I can also expose a perceived weakness to him without fear of repercussion, and his insight and perspective help me gain an objective view. He is not in the same field, but from that vantage point is ideal to read through a problematic section of a proposal to point out where a nonexpert will be unconvinced. Of course, I do the same for him. It is an amazing luxury to have the unconditional support of another scientist on call! L.C.B.

Beyond the advantages of having a partner also in science, husband-and-wife research teams can be brilliant. However, a potential drawback to such teams is that some colleagues and (worse) people with power to grant (or deny) funding, positions, promotions, tenure, etc. may assume that your contributions to the team were less significant than those of your partner.

> I am the first to emphasize that my husband, David Rogers, has been my greatest inspiration and I have benefited enormously from our joint work. When we were first married and our three children were young, we spent 15 highly productive years working together in Oxford and in the field in Africa on tsetse fly ecology. Soon after that period, my then head of department questioned the contribution I had made to our joint work. It had a devastating effect on me for some years, and nearly destroyed our team, until we decided to ignore such stupidity and carry on as before. S.E.R.

Of course, not all partnerships work to a fairy-tale design. The converse of the above examples is the partner who is not understanding or sympathetic of your scientific career demands. One who is unable to understand how you just cannot stop mid-sentence, mid-pipetting, mid-derivation, and come home for dinner! Supportive partners will ease the burden of our scientific careers, while unsupportive or overly demanding relationships can push the already high stress levels to the breaking point. This is why choosing the right partner (the supportive one) and working on good communication skills to ensure the relationship stays supportive is so critical to women in science. And that is important even before children!

Gender Equality in Choice–At Work and At Home?

As this is a book for women scientists, but one that we hope will attract some male readers, it is worth pausing to ask whether women face greater problems in balancing their professional and personal lives than men. While many men also work impossibly long hours, there is evidence that this comes less often at the expense of sacrificing a family, although many clearly sacrifice full involvement in family life.

Evidence of the extra challenge faced by women comes from the classic study on grant award statistics by the Swedish Medical Research Council (Wennerås and Wold, 1997) that suggests that women have to work harder (i.e., produce more top-class papers) just to survive. They uncovered an extraordinary sex bias in how the committee members matched their scores of the applicants' competence with their actual productivity. Amazingly, only at the highest level (quintile) of productivity were women applicants rated as highly as any of the men, but only as high as the lowest quintile of the male sample. A female scientist had to exceed a male colleague's scientific productivity by the equivalent of three extra papers in *Nature* or *Science*! No wonder women (with or without children) feel that, in the balance between work and personal life, they have no choice but to let work dominate.

Other studies show the cultural incompatibilities that women face. An analysis by Louisa Blackwell and Kevin Lynch (2002) of longitudinal data (1971–1991) held at the Office of National Statistics (U.K.) revealed that, for graduate women, but not men, in SET, continuing their careers was more at odds with full family life (partners and children) than for their colleagues in the health professions. In addition, Blackwell also cites qualitative work that suggests that incompatibilities between employment (of any kind) and family life are largely structurally determined and are weakening with the help of changing social and institutional arrangements. The quantitative census evidence, however, suggests that this may not be happening (fast enough) for women in SET occupations. She also points out that "women with degrees in SET, particularly the subjects where the representation of women is low, like mathematics, physics, engineering and technology, are exceptional. They survived the personal and social costs of atypical subject choices in education. It is plausible that they are more likely than other graduates to adjust their fertility behaviour rather than their career aspirations when confronted with employment practices that are incompatible with family life."

IDENTIFYING PRIORITIES. MAKING DECISIONS. ACHIEVING THE BALANCE YOU WANT

"What Is the Balance I Want to Achieve?"

The demands of our personal and professional lives will never reach a steady state; they'll remain in flux throughout our lives. Periodically, we'll need to reassess our priorities and work patterns, perhaps reining in for a time (i.e., the "punctuated equilibrium" described in "Climbing the Ladder"), in recognition of the value of duties outside the work place. Neither career paths nor lifestyles are set in stone. Perhaps with age, for example, the question of what you want to get out of your life (professional and personal) comes closer to the surface, giving you a clearer perspective on the value of an exclusive (obsessive?) focus on scientific objectives compared to the worth of investing in the next generation (be it through teaching, mentoring, raising your own family). It may inspire you to make choices that truly reflect your values and drive you to complete the important things before it's too late. But why wait until later to reflect on what will give your life meaning?

> *When it comes to overworking, it is useful to discover what your needs are and where you are getting them met. Think carefully about your needs (e.g. security, esteem, challenge, self respect, etc) and look at how many of them you are getting completely or largely through work. If you are to avoid the overworking trap you need to meet as many of your needs as possible by life outside work. In that way you can arrive at work full (and ready to concentrate on your job) instead of empty (and driven to meet your needs). Think of enjoyable and creative ways to meet your needs outside work.*
>
> Work Life Balance Center (2005)

The great thing about the scientific way of life is that, of all professions, it allows the greatest flexibility. It is innovative ideas that are in short supply that have the highest market value, and we can have ideas anywhere, anytime.

> *My contract reads: "Your hours of work are such as are reasonably required to carry out your duties to the satisfaction of your head of department. Should it be necessary for longer hours to be worked than those specified, no additional remuneration will be payable." Is this the ultimate open-ended contract?* S.E.R.

> *A friend of mine once said, "flex time is great–I can start work anytime before 7:30 a.m. and leave anytime after 6 p.m.!"* L.S.S.

Interpreted one way, such contracts could result in unreasonable expectations and measures of success being placed on an employee. It can, however, just as easily be viewed as an opportunity for an employee to discuss with her supervisor or manager the expectations and performance indicators of the position, ask for explicit clarification, and negotiate alternative ways of working that will fulfill both the employer's goals and the employee's desires or needs. For example, one research manager working for a large agricultural research firm in Canada wanted to leave work early each afternoon to be at home when her preteen children returned from school. She was able to demonstrate to her employer that her performance to date was exemplary (she consistently exceeded her targets) and proposed that she finish her day's work in the evenings. Having children had never compromised her performance to date, and this new arrangement would not change that, she argued. If anything, it would increase her productivity because it would reduce her stress levels. Her employer agreed to the proposal and paid for her to set up a home office (e.g., dedicated high-speed Internet access). She continues to exceed expectations at work and her own personal ones.

"I Want to Have Children. When Is the Best Time?"

No one need apologize for deciding that they really do not want children of their own, nor be made to feel they have achieved less—as women—than working women who have children. Similarly, those who know that they want a career in science *and* a family of their own need not apologize. Both lifestyles will be equally busy, equally challenging; it's just that the pressures, choices, decisions, and rewards will differ.

There will always appear to be good reasons why it is not the "right time" to have a baby. "We would really like to have children, but I can hardly keep up with my workload, let alone contemplate fitting children into it" is a common sentiment. But women in similar situations have done it before, and you can, too, if you truly want to have a family and are prepared to make the compromises that are a part of the decision. Within reason, never let work stand in the way of beginning your family, if that is what you really want. Ask yourself again "What do I value in life? What is important to me? What would give my life fullest meaning and satisfaction?" Once you decide that your life will include children, your life *will* change—forever—and you *will* balance your professional and personal lives differently because you have to.

But when is the *best* time to start a family? This is a question that no one can answer for you. Women achieve success in science whether they began their families as Ph.D. students or postdoctoral fellows, after a few years in their first jobs in industry or government, or after earning tenure. Most have the support of spouses or partners, others have adopted children and are raising them without partners, some succeed in spite of divorce—but all rely on their extended support network (e.g., extended family, friends, close colleagues). The considerations are different at each stage of your career (e.g., your time is more flexible when a Ph.D. student than as a research scientist in industry, but you'll have less financial security—and ability to pay for day care—than does an early career scientist). The only caveat is that of biology: there is no guarantee that you will be able to conceive, and the difficulty increases with age.

I really think the trend is in the right direction with respect to having children during a scientific career. From my anecdotal survey of women whom I have seen in academia, it seems that the first generation never had children at all, it was just not an option; the next generation waited until post-tenure, and then many of them had significant fertility issues, with a subset of those ending up like the first cohort; the next and current generation (and me) refused to wait for tenure and had children early during the first academic post. The rules have also been changing as we women have been pushing the boundaries—most top research universities in the U.S. have formal parental leave policies in place that not only guarantee time off when the baby arrives, but also delay of the tenure clock if desired. I was witness to and part of the committee that brought this policy into place at my current institution. This is real, tangible progress. L.C.B.

"What Is It I Really Want to Accomplish with My Career? What Kind of Time Do I Want to Spend with My Family?"

To determine how much time to spend on your career, you need to determine what it is you really want to accomplish with your career. This must be answered in the context of what else you want to achieve outside the workplace. The answers to these two questions can be in direct conflict with each other, creating tension and quite possibly guilt. To resolve the situation, you *may* ask yourself "What amount of time should I spend with people who depend on me, particularly my family? What amount of time should I spend on my career to achieve my aims?" These questions only exacerbate the problem because the word "should" itself implies that there is a right and a wrong answer with rules determined by some higher authority. (See strategies for defusing tension, eliminating "should," thinking clearly, and responding consciously—rather than reacting mindlessly—discussed in the chapter "Mental Toughness.") No one can answer these questions and certainly no one can answer them for *you*, and we aren't going to try. We can only reassure you that, once you have identified your priorities, the decisions about balancing the responsibilities and demands of family and career become much easier. In addition, the decisions that you need to make to support that priority also become easier. So, first get your priorities straight, then figure out the balance you want, and finally decide how to achieve it.

If only it were so easy! Priorities are constantly changing—with career opportunities, with the age of your children, even with the seasons. At times, they will seem clear, and that is great. Other times, they will be less clear. Have confidence. Do not give up hope! You will still be able to function, even as you struggle to identify what is best at the time and work toward it. As long as you can keep in mind your fundamental priorities, all the little decisions that you'll make everyday (e.g., "Will I agree to serve on committee X?") will become easier.

Even when you're certain that it's "family first" or "career first," you'll still face choices that beg the question "OK, but at what ratio?" This throws ambiguity back into the equation. As does the fact that deciding on your "ultimate" career goals is very difficult—even with the best information, predicting the future, and thus planning for career trajectories 10 or more years ahead is an exercise in uncertainty (as is discussed in the chapter "Career Management").

So our advice for those having difficulty deciding between competing priorities is to tackle hurdles one at a time, focus more on the more immediate question "How much time do I want to spend with my family this week/month/quarter?", and then schedule that time. It is easier to decide, for example, that going to all your daughter's soccer games and practices is important (and block off these times on your calendar at work), than it is to create a comprehensive plan of ranked priorities and make decisions between them on the fly. If you have blocked off all the soccer times, when you are approached to schedule another activity at work (e.g., special group meeting, student exam, guest seminar) at the same time, you can request that it be rescheduled, choose to miss the event, or decline the invitation. The career opportunities will come or not, but if you have worked within your priorities, you will have fewer conflicts and regrets.

Coming to terms with your priorities. I was complaining to a good friend one day about the intense emotional stress I was feeling about work and family issues. She said to me, "Linda, your problem is that you know what your priorities are, but you haven't come to terms with what that means for your career."

She was right: If I was comfortable with my priorities, then why was I feeling stressed? I did some soul searching and decided to take myself off the fast track. I had tenure, and it was time for my husband to build his career. He needed to travel a lot and, for our sanity and our marriage, someone had to be at home taking care of all those little things that keep a household running smoothly.

This was a very difficult decision for me. I had high expectations for myself. What I came to realize is that I could still achieve many of my fundamental goals—having a positive effect on students' lives and on science outreach programs.

The result? A 75% time academic position for two years with a reduced teaching load and more time to spend with my kids. I LOVED IT. It was the balance I needed. Did it take me off the fast track at work?—Yes. Did it end my career advancement?—No. Does it mean that I can't aspire to administrative positions in the future?—No.

Communicating those priorities. Once you have set your priorities, making them into a reality is also an art. I was not given a 75% time job. I advocated for it, planned for it, and achieved it. It was the result of careful conversations with my department chair and dean. I went to my department chair's office with a proposal—a plan of what I would do for 75% pay and the advantages for the department. It worked! L.S.S.

"What Am I Prepared to Give Up, and in Doing So, What Do I Gain?"

Once your priorities are set in motion, you'll have to face the consequences. Inevitably, there will be things you will have to give up. The cliché "you can't have it all" applies. You can't meet the school bus every day at 2:30 p.m. and expect your career to move at the same pace it would if you were working from 7 a.m. to 8 p.m. You simply cannot be a full-time, "fast-track" career woman and also enjoy the satisfaction of being the "classroom mom." You can choose many different points in the spectrum between those extremes, however. Whatever your decision point at a given time, there are adjustments that you will need to make to your personal and professional activities, some obvious and others subtler. Some fundamental choices are outlined below.

Work Part Time

One obvious option that we have already touched on in a different context is working part time. While this gives you more time with your family, it can mean giving up early promotion. People may question how serious you are about your career. You may be passed up for awards. If you are comfortable with your priorities, giving these things up may be uncomfortable but not devastating. And it is possible for this choice to be temporary and for you to move back into a faster track after some number of years.

Limit Travel

All the parents whom we talked to love to travel, but most have reduced it since starting their families. Doing so *will* limit you professionally, to a certain extent, by slowing your rate of advancement. To earn promotions, you need to develop a reputation that extends well beyond the boundaries of your local scientific community, and this necessarily requires travel. To be a world famous academician, for example, you need to give lectures around the world. And to build your research career in industry, you need to travel to meet with clients and attend conferences. The impact of decreased travel time can be minimized, however, if you choose your trips carefully, say, by traveling only when your presence is absolutely necessary to the advancement of your work and sending students or junior colleagues to represent you at less critical events.

Cate's Travel Tips

The "two night rule": Limit any trips away to a maximum of two nights, preferably weeknights, unless there are greatly extenuating circumstances.

Travel together as a family whenever possible. This can be done with a little extra effort. And is so wonderful when you have an infant and are nursing.

Combine work with vacation: the past several summers we have traveled for week-long trips to nice locations where we have both work and friends in the area. We bring the family, arrange for babysitting during the week, and get work done while still being with family. This has also provided great opportunities for the kids to see different areas and go hiking, swimming, or whatever with new friends in new places.

Let Go of Unrealistic Standards at Home

All of us would love to have everything in the house looking perfect and have regular time to exercise our creativity at home. But when we are balancing work and family, we need to accept some lack of perfection in the household. One approach to letting go of the pressure to keep the ideal home is to create "Ye Old Job Jar" (borrowed from the American comic strip *Hi and Lois*). The idea is to write on individual slips of paper the names of nonessential household tasks and projects that need to be done at some point and place them in a jar. Once stored away, the tasks stop cluttering up your mind and adding stress. And as you have time, tasks can removed, one-by-one, and completed. While this approach may be too rigid for some, it is a concrete way of recognizing that though there are many things we'd like to finish around the house, they simply are not a high priority until the children are grown and off to college.

You Must Learn to Say "No"

It is not always easy to act in accordance with your priorities. The demands from the outside world can be great and often are very subtle. So you need to learn to say "no" strategically, and with finality. *No running the PTA* (no matter how natural it may be for you). *No unnecessary travel for your job* (even though you may frustrate a colleague or two). *No becoming the soccer team's "mom"* (even if your best friends are asking). *No to whatever it is that prevents you from meeting your own priorities.* Every request that someone makes of you should be weighed in terms of your priorities. (The chapter "Time Stress" discusses strategies for setting boundaries and saying "no.")

> I became so good at saying "no" that younger people in my department would come to ask me how I had managed to set up my job to do all the things I like and none of the things I do not (yeah, right). What a reputation: "The NO lady." I'll let you in on a secret: I don't say "no." What I do is say "yes" to (or volunteer for) the things that I'd prefer to do before anyone asks me to do the things that I'd rather not. The department chair can't expect you to teach a course that you don't like, if you've already volunteered for one that you do. L.S.S.

PRACTICAL ADVICE FOR SCIENTISTS WITH CHILDREN

Work Environment

Like the real estate mantra, "location, location, location," you need to choose your work location as carefully as possible.

Geographic location is critically important, especially with respect to commuting issues. Commuting time is often wasted time when it comes to accomplishing any family goals, so any way to avoid it is worth considering. Both flexibility in working hours and location are important when dealing with unscheduled emergencies. The easier and faster it is to travel between work and home, the easier it is to take the children to doctors' appointments, deal with sick children, meet the plumber when the pipes are leaking, etc.

> Use that commuting time. Whether by car, by train, or by bicycle, my husband's commute takes 40 minutes. Therefore, he never drives. He either rides the train, where he can get "light work" done, or he cycles, which accomplishes his exercise for the day. This policy gives him extra time at work and with our family. L.C.B.

Pay close attention to your potential employer's policies regarding parental leave and flexible work arrangements, as well as the attitude among the employees regarding family issues. For example, are there many employees who are balancing the demands of work and family? Have any taken advantage of parental leave? If so, what has been their experience? Do they feel that they have the support of their colleagues?

> Flextime. One of the advantages of being a professor is what I call the "ultimate flex-time." The job is demanding and requires many night and weekend hours, but, except for the scheduled time of your classes (also negotiable with your

department head), the hours that you choose to work are your own. And where you work–at home or in the office–is partly up to you. Other jobs with this kind of flexibility may include certain types of consulting and freelance work. Take advantage of any flexibility your job allows! Consider the needs of your family, your work, and your coworkers and decide what the ideal arrangement would be for you and your employer. Then devise a plan, propose it to your supervisors, and push to see it implemented. L.S.S./L.C.B.

Division of Labor

Clarifying your relationship with your partner and understanding how each of you wants to interact with your children and contribute to the responsibilities of your home life is incredibly important. We are not advocating a particular division of labor but emphasizing the need to discuss the realities of a shared life with family and decide what is reasonable for your particular situation. There are myriad factors influencing how labor is divided, including the demands of each partner's job, work-related travel schedules, as well as the desire, willingness, and comfort level of each to take charge of the various responsibilities at home. For some couples, a 50/50 division is sensible. For many, it is not.

Take the initiative to talk through these issues with your partner ahead of time. Consider the number of hours each of your jobs requires (including commuting time). Discuss what you think is, overall, a fair ratio (work hours:children and household task hours) for each of you under your current circumstances and decide on an arrangement that you can both agree on. Be aware that this will need reassessing every year or any time there has been a significant change in your circumstances. For example, for a faculty member, the teaching load in a given quarter will change the ratio.

Deciding on the "ideal" percentage split between you and your partner is an excellent first step in the division of labor. You also need to decide roughly how to distribute duties, so that this percentage is achieved. And here we are at the "Who takes out the garbage?" stage. Don't skip this analysis—you needn't perform a survey with military exactness, but a rough estimate can be the key to avoiding real or even perceived inequities: Make a list of all the things that need to be done on a regular basis around the house and in family life and estimate the time required for each. Some items will be infrequent or seasonal (e.g., medical appointments, parent-teacher meetings, mowing the lawn, driving children to swimming lessons, soccer practices), but will often balance out, in the overall scheme. Now, divide the responsibilities. Work with it a few weeks and see how each of you feels!

Other Practical Tips

On Duty/Off Duty

It can be great relief to each partner to have one assigned day/night per week/month when he or she is "off duty" while the other is "on duty." The "off duty" parent can work, exercise, sleep, paint the house, shop, read...do whatever he or she desires. All family members understand that this time is 100% "free and clear," and the person who is "off duty" can pursue a chosen activity completely guilt free.

Alternate Who's in Charge

For many family activities, it is helpful to alternate which parent is in charge, especially for fun time with the children, like attending sport games and reading bedtime stories.

Babysitting Exchanges

A wonderful arrangement is exchanging babysitting duties with another couple (or couples) who has children whose ages are close to those of your own. You can supervise their children for an afternoon or evening on the weekend, for example (while they enjoy a movie, take a long bicycle ride, or enjoy a private meal out), and then, at a mutually convenient time, they can do the same for you. Sleepovers are even better.

Three-way (or More) Division of Labor

Many dual career couples have a third (or even fourth) person close at hand to help with some of the daily routine (e.g., grandparent, aunt, nanny). If you can, take advantage of this wonderful opportunity to share the load. Their contributions often extend well beyond the tangible; they are additional adult role models for the children, and, as their relationships with the children grow, they may be viewed as members of an "extended" family (even without a genetic link) and emotional support network.

Some couples have kept their nanny employed full time for many years beyond the preschool age of their children. She then becomes more of a "household assistant," who shops, cleans, mends, takes the children to doctor's appointments, etc., as well as supervises the children during after-school hours and school holidays. This is discussed further in the following sections.

Who's Watching the Children?

We've already touched on one of the major priorities to consider when balancing the responsibilities of work and family—attending to the needs of children—but it warrants further discussion. Issues surrounding high-quality supervision of children are probably the top concern of all working parents, and for women scientists with professional careers, it is critical. You have to know that your children are safe and well cared for in order to concentrate on your priorities at work. Obviously, you need the right child care. Of course, what is "right" for one family isn't the best solution for all. It is a very individual matter, and your needs may change over time. Here are some scenarios that you may wish to consider, in view of your own family's values, priorities, needs, and circumstances.

Member(s) of the Extended Family (e.g., Grandmother, Aunt, Cousin)

The traditional extended family arrangement can work extremely well but depends very much on individual personalities, personal circumstances, the nature of the preexisting familial relationships, etc. And it is much more difficult to expect things to be done according to your preferences if the arrangement is voluntary.

In-Home Care (e.g., Nanny or Au Pair)

There are many advantages to having paid, in-home care. It relieves the stress of having to get preschool aged children ready on time in the morning, as you can leave your children in PJ's eating breakfast as you dash out the door. More critically, when children become ill, they still will be cared for (unlike in day care), and the older ones will have supervision after school hours and when they're home on one of their (seemingly infinite) school holidays. Depending on your arrangement with your in-home help, the laundry may even be done and dinner started.

Hiring a nanny *is* expensive, and finding one whom you trust is an art. But many couples earning two incomes decide that the convenience and reliability of paid in-home support is well worth the expense. Even if it requires one person's full salary to achieve, this arrangement enables you to continue your research while your children are growing—for some, at a slower pace than if you did not have a family—so that, when child care responsibilities decrease, you *have* an established research program that you can ratchet up a notch, if you wish.

Another option is to hire an "international nanny" through a reputable au pair agency. This carefully selected person will live in your home and care for your children during working hours. The initial fee for applicant screening, matching to your family and needs, and making all the arrangements is substantial. But once she is living with your family, the extra pocket money that you give your au pair is small. You need a home that can accommodate an extra person with a separate bedroom, and you need to be comfortable making this new person a part of your family life. The cross-cultural exchange can be wonderful for everyone. One disadvantage is the length of your au pair's contract (12 months); you have to change au pairs every year.

Other Day Care

Taking your child to the home of a registered child minder or to a nursery can also work well, although it may not be as flexible as in-home care. Sharing with other families (as described in the previous section) can also provide daily playmates with no extra effort, but in both of these options, sick children must usually stay at home.

After School Care

Once your children are in school, you face the after school quandary: who will care for the children between the time they leave school and you return home from work? Fortunately, in most areas, there is a wide array of possible solutions, including programs run at the school itself and off-site after school programs. Grandparents, part-time babysitters, or even other parents in the same predicament may be glad to help out. Complex arrangements, for example, in which children have different activities each weekday after school, are positively relished by older children.

CHANGING ATTITUDES AND EXPECTATIONS

What Needs to Change? The System

While it is important for the individual to do what she can to be most effective in the system (as it is now), what is most needed is a transformation of the system. The "problem" of balancing professional and personal life needs to be turned around.

> *The ultimate solution lies not in women adjusting to a broken system, but in the system changing to accommodate the needs of a diverse work force.*

The emphasis must be on the "diverse work force"—not just on women—for improvements would benefit all, especially other underrepresented minorities. It is an unhealthy culture that expects creative professionals to work within rigid systems, irrespective of all the other calls on their lives (be they cultural practices, extended family responsibilities, less than perfect health, etc.).

As we have said, men need change as much as do women; the whole system needs to adjust. Encouragingly, this is already happening, largely due to the influence and efforts of women. Our gradually increasing presence and our practices born of necessity are changing attitudes, which in turn, will fuel more changes in practices. Now, more men *are* assuming greater responsibility for child care and participating more fully in the day-to-day work of family life.

We still need to correct the attitude (prevalent even amongst women) that reproduction is strictly a "female issue."

Some years ago, I was sitting on the University Staff Committee discussing the issue of maternity leave. New legislation to protect women's rights had raised the possibility of a backlash: fewer women might be hired as research assistants if there was a "risk" of their taking maternity leave. I made the point that it is not women who have babies, but couples; that it is something that we women do for men because men cannot do it for themselves. There was silence! S.E.R.

This is obviously a truism, not a revolutionary statement. How it gains general acceptance and brings about a systematic change is a topic for an entire book. Meanwhile, each of us can change the system, one person[1]

---

[1]For mothers, the following are some ideas to help push the system to becoming more humane:

Bring your baby to work. Put a crib and changing table in your office. Women all over academia are doing it. Even in 1979, the oldest of us (S.E.R.) did it very successfully while her baby was still immobile and slept most of the time. Students, both male and female, loved it and saw first-hand that work and family are compatible.

Take your baby on trips. Arrange for babysitting from the hotel or via networking with people in the area. Although it can be tedious to travel with an infant, at times, while the baby needs you (and you need the baby) every 3 hours, it may be at a time—not by marches on governments or conflicts with upper management—but by every woman making a difference in her own context. The beauty of this one-woman-at-a-time approach is that when you influence a system to adjust its outdated or unreasonable policies and practices, the women (and men) who follow you will expect the same consideration and will ask for more. And slowly, systemic change will occur.

Prove Your Worth to Gain Flexibility

"Success" means different things to different people. Whatever your expectations of work and life, rejecting other people's negative or unhelpful attitudes and replacing them with your own positive attitudes can enable you to define your own position in the workplace—but only if you do an excellent job *first*. It is unlikely that you will be able to walk into your first job and, three weeks later, ask for special schedules to fit your needs. You have to demonstrate the value you add to the organization and earn the respect of your colleagues before you request alternative work arrangements that would require them to make adjustments. Once you prove your worth, you will be in a position of strength, more able to convince them of the value of providing flexibility in the workplace and more likely to inspire change.

the best solution. From experience, we can tell you that a contented baby in a sling can certainly enliven a poster session at a meeting! It also challenges preconceived notions: people see you doing your job and fitting your family into it in creative ways.

Justify and request changes in your duties to accommodate your family needs. This does not have to mean requesting part-time status. It may mean setting up some telecommuting hours, so you can work from home or (for academics) requesting some administrative or organization duties in lieu of teaching one course. This would allow you more flexible working hours to accommodate children's piano lessons, sports teams, etc.

Never apologize for having responsibilities outside the workplace (children, elderly parents, etc.). If you do, you are unwittingly supporting the notion that only paid work is important and worthy of your time and attention. Commonly, mothers are so anxious not to provide any cause for continuing prejudices against women scientists, that they work doubly hard to fulfill their professional responsibilities.

The Signs of Change are Encouraging

Soon after I returned to the University of Oxford in 1974, to a six-year teaching post, I met David Rogers, a tenured university lecturer. Once our first child was born in 1979, my freedom to move to jobs outside Oxford (where tenured posts are in very limited supply) was restricted. I chose not to commute, and David preferred not to leave Oxford. My only career plan was "never, never, never give up."

In 1980, I was runner-up for a tenured post at Somerville, a college at Oxford. During dinner (a covert part of the selection procedure, when normally "off-limits" questions can be slipped into the conversation), one senior fellow asked me how many more children I planned to have. I said I didn't want my daughter to be an only child. She sniffed and said she didn't think women had the right to expect both a career and a family. How very different from a conversation I had with a fellow at another college a few years later: "Oh, you are so lucky. In my day, we had to choose between a career and a family. I chose a career and always regretted it. Now you can have both."

Freed (by default) of any teaching duties in Oxford, I won a research fellowship from the Leverhulme Trust, and we all went to Côte d'Ivoire (West Africa) for a 10-month sabbatical. I opted for a part-time salary, although studying tsetse fly ecology was a full-time job for both of us: we worked far into the evenings and often took Emily with us on fieldwork. When we returned to Oxford, I continued my part-time fellowship, while Thea and Jack arrived, but then I was awarded a Royal Society University Research Fellowship for up to 10 years. Even though Jack was only a year old, the opportunity was too good to pass up, so I returned to full-time work.

Neither we nor our wonderful child-minder wanted a full-time arrangement, so, with the support of our head of department, my husband and I instituted the "split day." Until Jack reached school age, for two days a week, we took it in turn to be at work (pre-Internet!) from 7 a.m. to 1 p.m. or 2 to 8 p.m. This may sound like a short day, but with no interruptions at either end of the day and no meal breaks, our productivity did not suffer. The other half of each day was packed with family activities and chores—they were actually extremely long days!

I have always packed my work around the children. They have grown up independent and loving—a far more significant achievement than any ephemeral professional performance indicator. They are the evidence, if I needed convincing, that our most valuable resource is not a pile of research papers but the next generation, whose education should have equal priority with research at universities. I have spent my professional life on an unbroken series of independent research fellowships, doing the science and teaching that I love. The stress of insecurity over 30 years has taken its toll, but I would never have swapped my family for any easier career path. I see encouraging signs that demonstrably capable women are now being valued and allowed the same rights as men to have the choice of both. S.E.R.

Every career has a "life." As you mature, you will probably make different choices regarding the emphasis that you'll place on research, leadership, and administrative involvements. You may ask yourself "Am I really the only person who can carry the burning scientific questions forward? Can I hand over the driving seat to my successors, recognizing that what they achieve will not negate what I have already contributed to my discipline?" You may feel you no longer need to compete or you may be able to ignore some of the pressures to publish a large number of papers each year. Relaxing a little and taking the opportunity to achieve different professional objectives could benefit us all: only respected senior practitioners, for example, are in a position to question, and have the power to influence a change in, the structure of the scientific academy.

ONWARD TO BALANCE

Since beginning the writing of this chapter, my priorities have changed dramatically and I have currently tipped the balance almost completely one way. I love being at home full time, but had we not moved following Charlotte's birth, I would (no doubt) have returned to work part time and enjoyed that balance just as much. If I have learned anything in the past 18 months with regard to child rearing—and it applies just as well to other responsibilities—it is that I need to do what works best for me, not what I think I ought to be doing, nor what I think others believe I ought to be doing. V.A.C.

It is maybe a bit ironic that I am working on this chapter on a Saturday morning in April. Where is the balance now? But as I think more about it, I realize that it is almost a perfect example of how I achieved balance. I'm awake, the children are asleep, the dog is fed, and it is pouring. My morning run has been canceled and so I am sitting down to work. I love my work and thus as I sit sipping my coffee, I am happy. L.S.S.

I love my professional career—my work in science, my service and outreach activities, my teaching; at the same time, my children are centrally important—nothing is more important than nurturing these young souls. Thus, every day I feel pulled in multiple directions, for in addition to family and work, I also have other activities I enjoy. My current strategy is scheduling—block out the times for a run, the times to be home early for piano lessons, and work fits in all around—early mornings, daytime, late evenings. And yes, right now, a sunny Saturday afternoon. My advice: find your balance, do excellent work, and don't be afraid to ask for flexibility to accommodate your personal life. For being happy in your career depends on being happy with your entire life. Good luck! Now, the lure of a bicycle ride is calling. L.C.B.

As I wandered round the Gothic castles of southern Bohemia (after attending a conference in Prague), I was struck by the obvious realization that the women portrayed on the walls not only had many servants but also had no professional work.

Pure leisure (how dull!). Among them was Elizabeth of Bohemia, noted as a woman with a very fine mind. How fortunate for women that they are now educationally fully emancipated. How fortunate for society that their intellectual powers are no longer wasted. How extremely fortunate for me personally that my life as a scientist has been enhanced by my full family life, which throws the crazy intensity of my work into its correct perspective. S.E.R.

REFERENCES

European Commission. *She Figures: Women and Science. Statistics and Indicators* (2003) [Online]. Available at: http://europa.eu.int/comm/research/science-society/pdf/she_figures_2003.pdf (accessed August 20, 2005).

Blackwell, L. and Lynch, K. *Women's Scientific Lives.* (Project report to the Economic and Social Research Council (ESRC) (2002) [Online]. Available at: http://www.regard.ac.uk (accessed through keyword search August 20, 2005).

Pearson, H. It's a scoop. *Nature* 2003;426:222–223.

Sanders, C. Analysis: science and the family don't mix. *The Times Higher Education Supplement* 2001;1508:6–7.

Wennerås C., and Wold, A. Nepotism and sexism in peer-review. *Nature* 1997;387:341–343.

Work Life Balance Centre [Online]. Available at: http://www.worklifebalancecentre.org (accessed August 20, 2005).

• •
MARGARET-ANN ARMOUR, Ph.D.
Associate Dean of Science (Diversity)
University of Alberta, CANADA
Chemist
• •

Dr. Margaret-Ann Armour is an internationally recognized expert in chemical safety and the disposal of hazardous wastes and an award-winning educator, mentor, and advocate for the promotion and advancement of girls and women in science. Born in Scotland and trained at the University of Edinburgh, she worked for five years as a research chemist in the paper-making industry before emigrating to Canada to complete her Ph.D. in physical organic chemistry at the University of Alberta. A member of the American Chemical Society, the American Association for the Advancement of Science, the New York Academy of Sciences, and a Fellow of the Chemical Institute of Canada, she currently sits on the Board of Directors of the Pacific Basin Consortium on Hazardous Wastes.

Many prestigious awards and honors have been bestowed on her, including the 3M Teaching Fellowship from the Society for Teaching and Learning in Higher Education, the McNeil Medal of the Royal Society of Canada, and the Governor General's Award in Commemoration of the Persons Case. The latter recognizes Dr. Armour's work in improving the situation of Canadian women in science and engineering and remembers the 1929 landmark decision of Canada's highest court to recognize women as "persons." In spite of all her achievements, Dr. Armour is eminently approachable and unassuming.

Her love of science and passion for teaching led naturally to a career in academe, where she has inspired generations of undergraduate and graduate students to pursue their dreams. As one of the founding members of WISEST (Women in Scholarship, Engineering, Science and Technology), an organization established over 20 years ago at the University of Alberta to address gender inequities in science, she has dedicated herself in particular to encouraging and supporting girls and young women. She grew up assuming that women can do anything they wish. Her hope is for all women to believe this.

● ●
DOROTHY TOVELL, Ph.D.
Alberta, CANADA
Biochemist
● ●

With a curiosity of spirit and a passion for exploring life in all its diversity, Dr. Dorothy Tovell is the quintessential "multitracker." Scientist, educator, and promoter of women in science, she has creatively combined several employment tracks in order to pursue her interests.

Her scientific career began at the University of Alberta (Canada), where she earned a Ph.D. in biochemistry. A postdoctoral research fellowship took her to the Central Public Health Laboratory in Helsinki and the National Institute for Medical Research in London to study interferon. An interest in virology inspired her to accept a research associate position at Queen's University (Canada), and after five years, she returned to her alma mater. She continued to study viruses for the next decade.

With time, she recognized that a life dedicated to research was not the best choice for her: she had neither the single-minded focus nor the drive required to be "successful," nor did the lifestyle suit her. In spite of the obvious uncertainty and insecurity, she resigned her position.

Since leaving full-time research, Dr. Tovell's life has become a wonderful tapestry of public involvements and private pursuits, woven together in a unique and ever-changing balance. She coordinated WISEST and chaired the Alberta Women's Science Network; each organization was awarded a prestigious science awareness prize during her tenure. For over a decade, she taught women's studies courses at the university. She continues to teach and lecture on topics of interest, especially science for the general public and autobiography for seniors, and provide health information to clients of her consulting business.

The nontraditional career path suits her well. Dr. Tovell views her many transitions as "bridges" to new lands of opportunity, discovery, and self-expression, and she finds satisfaction in this. "By my criteria, I am living a successful life."

Chapter

12

TRANSITIONS

Margaret-Ann Armour and Dorothy Tovell

> *Whether change chooses you or you choose change, it can still bring you new life.*
>
> J. E. Miller

Transitions, changes, promotions, setbacks, turning points, passages: these are inherent in our careers. Of course, specific patterns of change are individual. In addition, how we interpret and experience changes depends on our approach to career management or on the model that we choose for career development. Do we plan far ahead, setting stepwise goals for progress on our career path, or do we respond to opportunities that arise or inner shifts in commitment, seeing our career as a meandering journey through our life? Is our dream a successful pursuit of an occupation, one that advances our status, rank, salary, and reputation? Are we in it for the long haul, or do we prefer short-term commitments? How important is it that we make a contribution to science, and what do we expect in return?

Many of these questions have been addressed in the chapter "Career Management," but we believe that it may be useful, even necessary, to revisit these issues when thinking about transitions. Climbing the ladder, the topic of another chapter, is also related: how we make sense of moving from one rung to the next, remaining in place or even slipping back. Might we decide to step off the ladder altogether? Transitions are also very much related to balancing our professional and personal lives.

In writing this chapter, we focus on stories from women scientists rather than on specific strategies for navigating change. We invited a number of colleagues to send us their stories of a transition (indicating that they would remain anonymous); their contributions are included. We begin, however, by introducing a model that explores the nature of "change" and "transition." The terms are often used interchangeably, but William Bridges distinguishes between the two: "Change is situational: the new site, the new boss, the new team, the new policy. Transition is the psychological process people go through to come to terms with the new situation. Change is external, transition is internal...Unless transition occurs, change will not work" (Bridges, 1991, pp. 3–4).

How easily we make the internal transition inspired by a new situation may well be related to how the external change is initiated. Some changes we choose, others are thrust upon us. If we consider traditional career paths within an institution or industry, there are changes in which we make the expected next move, changes that are explicitly imposed, changes that we are expected to make but that are not explicitly stated, and changes that we are not expected to make. In the experience of one professor of engineering:

> In the first category, "empowered change," I put the promotion from high school to university, from university to graduate school, from graduate school to professor, and from tenured professor to full professor; also marriage and birth of first child. In all these cases, I was well prepared for the change at the time when the change was expected to occur. The path from one step to the next was clear, and the expected outcome of the change was positive. All these changes were well supported and were effortless.
>
> An example of "imposed change" is the change that can occur in institutional values, for example, from a balanced focus on undergraduate and graduate education to becoming a research-intensive institution. The request for change is explicit. The members of the organization can embrace the change, resist the change peacefully (Gandhi model), or react to the change by leaving or running for an opposing administrative position (traditional, male, "my way or the highway" model). The first approach makes the requested change an "empowered change"; the second moves the individual into an extended transition phase where the outcome is not immediately clear; and the third option precipitates a new change cycle.
>
> I emphasize that both models of resistance may or may not have positive outcomes. Most successful people use quiet resistance as one of many strategies for dealing with change (also known as "choosing your battles"). Dealing with imposed change is the essence of the strategic dance of power that is so difficult for the novice; the advice of mentors and friends is almost essential to avoid

errors in judgment. In my experience, the most difficult situation to deal with is when others expect us to make a change, but the request is not explicit and the criteria for success are transparent only to those who expect the change to occur.

If "empowered change" is like riding a wave and "imposed change" is navigating rapids, then "expected change" is akin to meeting an iceberg in the fog. And like the Titanic, even the best prepared, beautiful, intelligent, tough, charming women (and men) have been sunk by these invisible hazards.

The final category—changes that we are "not supposed" to make—includes changes that simply are not encouraged, as well as ones that may be interpreted as failures (by others or by ourselves). We are much less likely to receive support from those committed to "the system" for a decision to concentrate on teaching (at the expense of research) in a research-intensive institution or for a change in plans resulting from denial of tenure. Later in this chapter we provide examples of how taking a step backward to move forward is sometimes the right thing to do: one woman left a research environment to move to a teaching institution and another moved laterally, from marketing coordinator back to a design engineer.

Many of the changes that we have been describing were externally initiated, such as the predictable stages in our education, moving from one rank to another in academe or industry, adjusting to new work responsibilities because of a decision made by a supervisor, or having to relocate and begin a new job because a partner has accepted a job in a different community. The impetus for change can also come from within, for example, from a gradual or sudden realization that one's work is no longer satisfying or that a different kind of work is very appealing. One woman trained in biochemistry describes her experience:

Gradually, I came to realize that I did not have the intense personal curiosity about how viruses replicate to provide sufficient motivation for me to spend my working life in research. I much preferred to teach, to pass on information that originated with research done by others. Later I became more intrigued by questions about how science is done than by science itself, so I moved on to teach a course in women's studies that included feminist critiques of science.

Not only did this woman realize that her work was no longer satisfying, but also that there were viable alternatives—an important next step in the process. (Later in this chapter we present a similar example, of a woman who thought about her skills, knowledge, and attitudes and realized that she had the ability to be a director.) A researcher in an academic institution may stop working on a research problem and move to a new topic of investigation because she truly has lost interest in the old project and/or is fascinated with the new one. Or the decision can be externally triggered, by changes in availability of research funds, for example. Indeed, it is quite likely that many changes will be initiated by a number of factors; real life seldom is as simple as our models.

It is clear from these examples that transition does not always follow change but may, in fact, precede it. The researcher who tires of her project but finds a different question very appealing has begun the inner process of transition before she makes the change in her work, whereas the one who switches to a new project because she cannot obtain funds to carry on her current project may well have to come to terms with that change after she makes it. Bridges' model suggests that the change is not likely to be successful until she makes a transition—until she transfers her interest in research questions on her old topic to questions on her new one.

The process of transition is less obvious than a change—at least from an observer's point of view—and usually takes much longer. While some transitions occur almost overnight, many transitions are so slow as to be imperceptible; only in looking back do we realize that we have moved. In Bridges' model, a transition consists of three seemingly paradoxically defined stages: the beginning (which is really an ending), middle (which he calls a "neutral zone"), and the end (which is really a new beginning). Where change focuses on outcome (a new beginning), transition begins with an ending (letting go of the old): one must leave the old situation and *attitudes* behind. Most often our difficulty with transition is in expecting to begin with the outcome. Bridges believes that it is the letting go that we resist most and strongly suggests that in any time of change, we examine what we might have to give up. Sometimes we may realize that we are not yet ready to let go, and this could mean declining an opportunity for promotion. One chemist explains:

> I had done the same kind of work in quality control and health and safety for a drug company for a number of years and was talking about being ready for a change, when my supervisor offered me a new position as a clinical monitor. It would mean that I would have to begin the new position within a very short period of time. I was concerned about the continuation of my current work, but no one could assure me that someone else would assume the responsibilities. I turned down the new position because of this, and because I simply could not accept that it was "not my problem."

In this case, a change in job did not occur, although the process of transition had begun, since the woman was aware of a diminishing satisfaction with her current work in quality control and a growing desire to assume new responsibilities. But she hadn't quite given up her old attitudes or established new ones. Bridges (2001) explains: "In between the letting go and the taking hold again, there is a chaotic but potentially creative 'neutral zone' when things aren't the old way but they aren't really a new way either" (p. 2). We need to appreciate both how uncomfortable this can be and how much opportunity for creativity it gives us. A scientist who had enjoyed being a successful researcher wrote:

> The "neutral zone" described by Bridges is apt in terms of my determination to leave my position and the unsettling situation that ensued in subsequent months. Science has always been my passion, and I wanted to explore ways of using my talents in alternate ways, in areas of employment in which I would find contentment and security. But it was a struggle to let go of the old career expectations that were based on my years of training. I came to terms with the fact that this would be a period of learning—for both my children and me—and that life as we knew it would not be the same for some time.

Whether triggered by specific external changes or by inner processes, transitions are characterized by the same three stages: the ending (letting go of the old), the neutral zone, and the new beginning (embracing the new). These stages are seldom as clear-cut or sequential as the simple model may suggest; there may be some overlap, some back and forth movement, or periods of no movement at all. We need to acknowledge the process and be patient with ourselves as it unfolds.

During transitions, we may be surprised by the strength of our emotions. In letting go, for example, we may experience feelings of loss that may also be associated with strong feelings of grief or fear of the unknown. The neutral zone may be characterized by feelings of irritation, frustration, or depression. Joy and hope may accompany new beginnings. Or we may feel very little. Regardless, it is wise to take especially good care of ourselves during times of change and transition; to cultivate healthy eating, sleeping, and exercising habits; and not to be afraid to call on others for practical or emotional support when we need it (Miller, 1997). We may be fortunate enough to have supervisors or managers who understand the process of transition and can offer specific suggestions and strategies that are relevant to our situation. Or we may find encouragement through reading books such as *Managing Transitions* (Bridges, 1991).

Some transitions we make more consciously than others. As long as we live and work, however, there will be changes and their accompanying transitions to challenge us. The more we understand the process and our own transitions, the more we can learn from them and the easier they may become.

CHANGES THROUGHOUT OUR LIVES

We now consider some points in the development of a career in which changes occur and illustrate them with real-life stories. Many, but not all, of these changes can easily be anticipated and are clearly laid out within traditional systems. You'll see that we comment sparingly on the stories; preferring to let them speak for themselves and leave their interpretation to you, in the light of your own experiences and Bridges' model of change and transition. Perhaps you already have made some of the changes or transitions yourself; you may find it useful, then, to reflect on them and, as a teacher, mother, colleague, or mentor, share with others your own thoughts and those of the women who tell their stories in this chapter. The experiences of women who have been able to thrive within the current system can provide a guide to women entering it, women who (we hope) will eventually influence and change the system itself—to the benefit of all.

School to Postsecondary Education

The transition from being a high school student (enjoying and doing well in science classes) to succeeding in a science program at a postsecondary institution is a major one. Working beside other students who also enjoy science is exciting and has many rewards. But there is a shift in the approach to teaching and learning that first years, especially in their first term, often do not anticipate: they must let go of the high school expectation that they will "be taught" and recognize and accept that now they have to take responsibility for their own learning. Professors often are aware of the potential difficulty of this transition and allow students leeway in first term. The letting go process often happens during this time—a lesson poignantly learned through midterm examinations failed and an impossible study schedule for the last week before final examinations. By second term, the transition is made to a different way of learning, and the year is salvaged.

I came very close to failing my introductory physics course at university many years ago. My professor suggested that we do, for example, questions 2, 3, 5, 6, 7, and 9, and hand in questions 11 and 12. I didn't take his suggestions seriously, and attempted to answer only questions 11 and 12, usually with limited success. Yes, I was now a university student, but I still held attitudes applicable to high school; I expected that the teacher would lead me through the whole process of solving specific problems, first the simple and then the more complex. I had not realized that I was now responsible for my own learning, and I came very close to not making the change to a successful student.

Being prepared for the transition before arriving at a university makes the first term much less traumatic and decreases the likelihood that new students will choose to leave their studies or change to a different field after the first year.

When a young woman has to leave home and move to a new community to continue postsecondary education, another kind of transition has to occur. A rural student registering in a large urban university or college, for example, may discover that the number of students in her first-year science classes is greater than the population of her home community. How easy (and natural) it would be for a somewhat introverted young woman to feel overwhelmed by the sheer numbers of people, isolated and out of place in this strange mass of mostly teenagers, where everyone else seems to have a group of friends.

A change in situation such as this requires giving up the security and support of home and family. It takes time. Yet the new exciting world of young friends with similar interests can soon ease the transition. Living in a residence is often the better way of managing this transition, rather than living alone. The lack of privacy may be difficult at first, but at least it offers the company of people (often in the same courses) who are dealing with the same experience of moving away from home and with whom you can share difficulties and solutions. Through these shared experiences, lifelong friendships can develop. Yet during this transition, there can be a vague (or sometimes strong) sense that you are missing information to enable you to make good decisions, even with many student friends, but you do not know the questions to ask. You are uncertain, for example, that your peers have the appropriate knowledge of your particular program to give you the direction that you need. Rest assured—all postsecondary institutions have academic advisors and personal counselors who have listening ears and the information, knowledge, and experience to offer support. Seek them out and consider their advice seriously.

Postsecondary Education to Employment

You have completed your science diploma or degree at a postsecondary institution: now what? During your studies, you may have discovered a science subject about which you have a passionate interest. Or you may be thinking "I like science, but are there jobs in my current field?" The answer may be relatively clear if you attended an institute of technology (where your training prepares you for a specific type of employment) or pursued a professional program of study. But it may be more complex if you've earned a general science degree. Finding a compatible job is one part of the transition from education to work and is discussed in the chapter "Career

Management." There is another aspect to this transition: moving on from having to learn a prescribed body of knowledge on which you will be examined to understanding and fulfilling the expectations of your new workplace. How different will they be? You will still have a supervisor, mentor, or manager whose responsibility it is to define the expectations of the position, but, like moving from high school to postsecondary studies, you will need to be able to recognize when the job can be done independently and when you need to question and clarify your responsibilities. It is time to move beyond learning a defined body of knowledge to developing an exciting new ability: taking initiative and making suggestions about the work that you are entrusted to do.

This transition may also be one of moving to independence. If you have lived at home during your postsecondary studies, now may be the time that you have to move away to find appropriate work: What if your family does not approve of your choice? In this case, moving away from security and support becomes a more traumatic transition. A zoologist describes her experience:

> In university, after taking a few zoology courses, I developed a love for the discipline. In fact, I loved it so much that, in my third year, I accepted a field research assistant position. My parents certainly did not approve of their daughter working in a bush camp. In fact, some of my relatives even commented on why I would do such things. This kind of work was completely foreign to them and seen as unacceptable and of low status. You see, they wanted me to go into medicine, not go "loafing around in the bush chasing some animals." But my passion for zoology was only further reinforced after my first field season. This was a very difficult time because none of my family members supported or even understood this passion of mine. They saw it as a phase and not something a respectable girl would do. After all, what decent boy would want to marry a girl who worked in the bush? I felt very alone. I soon realized that this was not a one-time occurrence, but rather, the beginning of taking charge of my life and making my own career choices and decisions.
>
> There would have been nothing wrong with following the dreams and wishes of my family—if they had been my dreams and wishes as well. But this was not the case! I knew I had to find out who I was as a person. So, I journeyed to other countries, alone and with friends. I joined an international association on campus and worked abroad. I played in a hand bell choir and participated in many volunteer groups in our community. I did many more field seasons of zoology research. Ultimately, I explored the world around me, not just in an academic sense but also a personal and spiritual one.

"My passion for zoology"—what a wonderful way to describe the reason for making the transition to a mature and independent young woman, even when this required letting go of family expectations. A passion for our scientific discipline allows us to take flight from the foundation of our education to do work that is fulfilling.

Postsecondary Education or Employment to Postgraduate Study

One of the concerns expressed by science administrators, policymakers, and analysts is that women are not moving into decision-making roles in the sciences at the rate that one might expect from the numbers registering in undergraduate science programs. Completion of the bachelor's degree is one of the junctions at which young women consider very carefully many factors when making the next career choice, not least of which is the possibility of combining postgraduate studies with family life. They are very much aware that the biological clock is beginning to tick and ask many excellent questions, such as: "How does my partner's education or career affect where I go to graduate school?" "If I choose to do a Ph.D., when will I be able to have a family and how will I manage a family and the intense research demands?" "If I opt to go on into a master's program, will I be qualified for the kind of work that I would like to do?"

In addition, choosing to attend graduate school may mean yet another change that requires a move away from a known environment, even leaving one's own country. With all the excitement that this brings, letting go of the known and moving on to the unknown still can be painful. This story, from a chemist who currently works as an academic administrator, is a good example:

One of the biggest transitions in my life was going to graduate school in California. Not only did I change countries and continents, I changed cultures and time zones. California in the 1960s was eons apart from the isolated islands of New Zealand. Nowadays, frequent and affordable airline flights, cheap international phone rates, and the ubiquitous Internet make global travel and communication no big deal. But it was a very big issue when I arrived in California. Home was a long way off; it had been very expensive to buy my one-way air ticket and I did not have the money for a return fare. I could not think of returning at Christmas or for a holiday; trans-Pacific telephone calls were so expensive that they could only be made in case of emergency and so airmail letters that took about 10 days were the only way of keeping in touch with friends and family. Adjusting to the many differences took time. I remember that first semester seemed so long, but by the second semester my confidence was returning. I had made some friends, was starting to understand the system, and realized that I could indeed cope with graduate school. Not only that, I was enjoying it! It is only on looking back that I recognize how profound the transition really was. At the time, a mixture of youthful naïveté, blissful unawareness, and a perceived need for something different propelled me along this path. Today, I wonder how and

> why I did it; at the time, it never occurred to me to question why I was setting out on this great adventure. I have never had any regrets about the choice I made.

Although this kind of move is now less isolating, attending graduate school still is a major change. It requires making the transition from formal course learning to participating in the discovery of new knowledge; from knowing that there are solutions to undergraduate problems, to needing to think of new questions and guessing whether the answers are anywhere within the reach of persistent hard work. Graduate school is an apprenticeship—the bridge between being an undergraduate and a research director in industry or academe.

But what about the transition from paid employment to postgraduate study? Perhaps you have found that the kind of work that you see your colleagues doing and that you believe you would enjoy requires a higher level of education than you currently possess, so you have made a conscious choice to return to an educational institution for further study. One university administrator recalls what the transition from work to postgraduate study was like for her:

> After five years as a chemist in the research department of a Scottish paper mill, I decided to study for a Ph.D. at a Canadian university. There were placement examinations on arrival, and, although I had reviewed my undergraduate course material, five years is long enough to forget facts that have not been used. I found myself having to retake courses and questioning whether I would succeed. This was devastating; I had come half way around the world to follow my dream and I might not make it. By half way through the first term, I rediscovered how to study and my confidence was returning. At the end of that term, I chose a research director, became part of a research group, and thrived.
>
> What helped me over that bumpy transition? The graduate student advisor was a kind, approachable person who helped me to realize that I was not unique in finding it difficult to return to academic study and suggested I give myself time to adjust. My research supervisor had high expectations of his students; that was exactly what I needed. I could believe in myself again. I respected him and appreciated the kind of guidance he gave and independence he encouraged. My Ph.D. defined a path for an amazing journey, and I shudder to think I might have given up during that initial bumpy transition.

A recurring theme in these stories of transitions is the importance of seeking out people with whom we can share our successes and difficulties, people whom we trust and respect and who have the appropriate knowledge and experience.

Postgraduate Education to Employment Inside Academe

You have spent time as a postdoctoral fellow and now are ready to move from supervisor-guided research to being an independent researcher. This is another transition in which we can move forward in self-confidence and level of maturity; we can be honest, realistic, and assertive about our experience and qualifications. Ironically, women generally tend to underestimate their abilities more than men, making the transition to an academic position in a high-powered research university, for example, more difficult. In addition, workplace values and practices tend to be less comfortable for women than for men, and the measures of success are traditional. And in some disciplines, such as the physical sciences, the percentage of female scientists is low. Yet the rewards of a career investigating problems that you can more or less choose for yourself and whose solutions may benefit people are very high. A research chemist remembers a definitive moment:

> That morning, sitting at the coffee table with 10 members of my research group, is etched in my mind. It was early in my career; my research had just been funded and I was working with a group of wonderful people: young, excited students, and older, experienced researchers, who all were interested in working on this project. The realization dawned on me of how privileged I was to be an integral part of this laboratory family, to share my passion with them for the work that we were doing, and my own excitement rose as I sensed the possibilities for the future. My transition to an independent researcher had been made.

Postgraduate Education to Employment Outside Academe

You have just completed a Ph.D. and perhaps also a postdoctoral fellowship. Now it is time to take all these scholarly qualifications and use them for the benefit of society. You are making a transition from discovery of new knowledge (which is largely curiosity based) to goal-directed discovery of processes and knowledge that will lead to "better" products. The transition to a first, nonacademic position can be influenced by nagging feelings of inadequacy: "Will I fulfill the expectations of my employer?" you ask yourself. At least in academe, you knew what was valued and how you would be judged; you had a supervisor who laid out the expectations of your fellowship and, perhaps, you had a supportive research team within which you worked. You had always heard that the "real" world was very different. Now you are going to find out whether this is indeed true.

Everyone experiences these feelings, though they are particularly problematic for women. We tend to underestimate ourselves, and this sometimes results in our not even applying for positions in which we're very interested because we judge (incorrectly) that we are not qualified, simply from reading a list of job requirements. But our response to these feelings is a resounding: "Yes! You *will* succeed!"

Depending on the nature of your graduate research project and experience of graduate school, the transition to work may be a very smooth one. If, for example, graduate school was not a very positive experience, then the transition to work can be an especially happy one. A biologist who enjoys working and teaching in health-related areas tells her story:

> The transition from my Ph.D. program to my work outside academe occurred about 18 months ago. More accurately, the external change in life status from graduate student to highly qualified and well-paid employee occurred at that time. The actual transition—the shift within myself—was a much longer process. It is said that letting go of the old is the hardest part of a transition. I found quite the opposite. I was happy to leave my life and persona as a "grad student." My experiences in graduate school were not the happiest, although I value the strength of character that I discovered in myself through the process.
>
> The difficult part for me was figuring out who to be next in this new life. It was fascinating to realize how much my changing perception of myself was influenced by how others perceived me. In graduate school, I felt very low in the pecking order. My opinions did not count for much; my words and actions had only very minor implications beyond myself; much of the external feedback I received from those in immediate control of my destiny seemed to suggest that most of my professional contributions were minor and of no great interest. These messages were at odds with my own internal belief in my work and my personal and professional abilities. However, over time, that inner voice became quieter and quieter, drowned out by the constant battering from without.
>
> When I entered the workforce as a newly minted Ph.D., the external feedback changed drastically. I had gone from the bottom of the old pile to the top of a new pile. Although I was still the same person, societal perceptions of my professional abilities altered completely: I was no longer a graduate student; I was a "doctor," an "expert." Truthfully, the respect that I received from new coworkers made me uncomfortable for some time. The urge to look over my shoulder, to see whom people were talking to with so much respect, was strong. I was so unaccustomed to having my voice heard, my opinions requested, respected, and valued. Slowly, day by day, the inner voice that had always believed in me grew stronger. I was accustomed to being invisible and unheard, hidden in my work; now I was expected to take a prominent, visible role, both with my coworkers

and in a much more public venue. I began to grow, to transition into the person I needed to become: strong and vocal, someone whose opinions count, whose actions and words have profound and far-reaching implications, respected by her colleagues and by those who meet with her for her professional expertise as well as for her personal qualities. My inner voice has become strong once again. It says: "I can succeed!"

Work to Full Time Motherhood and Back to Work

You have extensive academic qualifications in a scientific discipline that you thoroughly enjoy, but the choice about when to start a family has to be made. You conclude that having a child should not be postponed, but you know that you want to continue to pursue your profession. You know that the demands of your personal and professional life will be greater than ever, and you'll need patience and skill to balance them in a way that works best for you (as you read in the chapter "Balancing Professional and Personal Life"). You face many conflicting options and make difficult decisions; just make sure you've chosen a supportive partner!

The break from work to have a family may be short or prolonged. Often, two transitions are involved: leaving study or work to have a child and returning to work some time after the baby is born. In North America, going on maternity (or paternity) leave has become legitimized, while taking breaks for other reasons may still be suspected as indications of a lack of commitment. Companies are making it as easy as possible for young mothers to return to work: day-care facilities are available, and sometimes, part-time or flexible work hours can be negotiated. Some granting agencies (e.g., in Canada) have developed funding application forms that provide a section in which research scientists, returning to full-time work at a university, can explain their absence.

But this does not ease the guilt that many women feel when they leave their baby with someone else, so that they can return to work, nor does it compensate for the sense of loss at missing parts of their child's development. Whether we return to work out of financial necessity or because we love our work and simply cannot imagine not being able to practice in our chosen field, we may experience these emotions. This is a time when supportive female colleagues who have had the same experience need to be sought out, for they can ease the transition. A senior industrial chemist reflects on going back to work after her child was born:

One of the transitions in my life that I found to be very challenging involved becoming a mother and then returning to the workforce. Many friends and colleagues hinted that I would become bored and frustrated at home with a baby, but in fact, I found it to be quite the opposite. I left my position as a postdoctoral fellow a few months before my son was born, since we were moving to another city and buying our first house. I thoroughly enjoyed that summer as we settled into our house, and I was fortunate to have many friends who were also starting their families around that time. After my son was born, I was completely occupied with his care and, like any new parents, my husband and I were fascinated watching his development, his first smiles, steps, words, etc. I discovered that, for me, there is no research project that could ever be as engaging as your own infant.

Although I didn't really miss my job, I did know that I wanted to return, and I suspected it would be difficult to reenter the scientific workforce after a long absence. I was very fortunate to find a part-time job, three days a week in a small lab. But I was not at all prepared for the emotions that I felt on leaving my six-month-old son. Even though I had an interesting job with great people, I found it very difficult. Working part time and having wonderful neighbors and friends to babysit made it manageable, but I'll admit that I felt very "divided" at times. Eventually we ended up hiring a live-in nanny. That helped a lot when I chose to change jobs and work full time with a major multinational chemical company where I have now worked for 21 years.

I found things did become easier over time, but I think most mothers continue to find it very challenging to strike a good balance between their work and their family life. One of the things that can really help through these transitions is, first and foremost, a supportive spouse who is willing to pitch in and help. With dual careers, both spouses have to be willing to juggle careers with appointments, helping out at school, staying home with a sick child. I've been very lucky in that regard. The payoff for me has been tremendous and I can't imagine how my life would have been without being able to enjoy my career and my family.

After 10 Years in the Workforce

After about 10 years in science, some take a critical look at their lifestyle, goals, and experiences and make new choices for their lives. Women, in particular, want to work in environments where they feel valued, where relationships are important, and where there is a supportive community. They may decide, on reflection, that their workplace no longer meets their needs and choose to change the direction of their work, or even opt out of the sciences altogether. It may be that the work situation has become so untenable that the decision to move on is not a difficult one.

Or it may be that they sense that it is time to look for new challenges to fulfill their own ambitions. In either case, letting go of a secure job and known lifestyle to step into the unknown requires risk taking and courage—and can take time and conscious effort. When a woman is a single parent with responsibilities to her children, deciding to move can be even more difficult.

Of course, there are positive, energizing aspects to making the decision to change. The excitement of a new job, the sense of being able to use one's expertise even more effectively than previously, and the recognition that this is your own choice and that you have taken charge of your life all are motivating, as a senior university teacher and administrator attests:

> The transition that I would like to share involves my departure from a research environment at a university in Canada to pursue new directions at a teaching institution. As a postdoctoral trainee and later a research associate, my perception of the Canadian research environment was fairly naïve. Being a graduate from Australia, as well as a single parent of two young boys, I struggled with the expectation of working strange hours in the laboratory and dealing with the "publish or perish" mantra that is so inborn in North American institutions. I also had to cope with the cultural changes that beset my children and me in this new country. I immensely enjoyed the bench work, was adept at investigating the various aspects of the work, and achieved the goals expected of me during my training. I also appreciated the interactions with my peers during my postdoctoral studies. However, after several years of work at another laboratory within the university, I was unable to abide the negative atmosphere created by some individuals; I had never previously experienced such unacceptable behavior. Despite many efforts to resolve the situation, it became obvious that it was irreparable.
>
> I have never been a quitter, but, to be quite honest, I knew that if I remained in this environment, it would destroy my inner well-being as well as damage my relationship with my children. Having to make the conscious decision to leave the situation was one of the most difficult choices of my life. I had to weigh the security of my children and my fear of setting out in uncharted territories against the consequences of staying.
>
> The new beginning in my career came about rather suddenly: I was still working at the university when I received a call in response to a letter of introduction that I had circulated to several establishments. Now eight years into my new position, I can honestly say that my work is enjoyable and satisfying. I can actually work in areas that I am passionate about. I interact with students in advancing science in interesting ways. I share outreach and promotion of girls in science with women who are champions in this country. I make positive contributions to the university and general community. I have a fulfilling relationship with my family. Making the scary jump from a situation that I feared to "let go" of to the work I now perform is liberating and nurturing. I have had

opportunities to grow, not only as an educator and researcher, but also as a human being who can give something back to her family and the community. It is this transition that I will forever value as my career continues to evolve over the years.

Mid-Career Changes: Promotions, Lateral Moves, Change of Profession

As mentioned earlier, a current concern among science administrators, policymakers, and analysts is the lack of representation of women and other diverse groups in decision-making roles in the sciences and engineering. But what kind of transitions do women in science experience as they move into more senior positions? What are some characteristics of a transition into management? Early in this chapter, we discussed the change in approach to learning that is required in the move from high school to post-secondary education and postulated that knowing about this ahead of time can be helpful. The value of knowing what to expect in a move to a more responsible position later in a career is equally true. Three women share their very different experiences. The first one, a director of technology for a school board, reemphasizes the importance of being a member of a supportive team and of recognizing one's own ability: "I realized I had the skills to be a director."

When one begins a new endeavor, one is excited about the prospect, yet at the same time is apprehensive. One begins to question whether this is a good idea. Such is the case for many of us as we move from one position to another. And such was the case when I became director of technology services for a school board. William Bridges (1991), in Managing Transitions, suggests that people need "the four P's"—a purpose, picture, plan, and part to play—to make a new beginning. I was fortunate to have a supportive administrative team to help me with these. The superintendent at the time had a plan and was able to provide me with a picture of what the organization expected. He was able to clarify and communicate the purpose. He provided a picture of the vision the organization had for technology and my part in it. Central to making the transition was the fact that key administrative members realized the importance of my being there.

When I reflect on this period, I realize that it was an important time for me to take stock. I remember thinking about my skills, knowledge, and attitudes—what Bridges defines as "resources," potential advantages that all of us have—and that I realized I had a variety of resources to offer! I had the skills to be a director.

Moving from one type of work into another meant many things to me, but it also meant letting go. It is difficult to let go of things that are so comfortable. Just as new shoes don't fit like the old ones, moving from the comfort of one job into another resulted in some pressure points. I believe that I was able to deal with these because my science background provided me with the ability to experiment. Bridges writes about the importance of experimenting a little each day. He notes that doing so forces us to do things differently. I challenge each of you to experiment each day. Look for ways to experiment! Be willing to ask questions, be willing to set goals, be willing to take a risk!

In the second story, a senior government administrator shares the importance of establishing authority and earning the respect of those who work for you, each time a move is made to a different management position.

I found out very quickly that I have to start all over again, in each new workplace, to establish my authority as a senior manager and my credibility beyond the scientific credibility of my doctorate degree. Junior professionals, particularly women who do not hold advanced degrees, view me as "one of them." Often, they refuse to accept that I, being in a senior leadership position, do have the authority to anticipate problems and make decisions. My transition involved accepting that I must work that much harder to clarify roles and responsibilities and establish my credibility and authority in each organization that I join.

The third story, from an engineering professor, illustrates the importance of open communication with those in authority so that expectations are made known and can be discussed.

When I became an associate chair responsible for the undergraduate program in our department, I recruited a team of advisors who were very dependable. Like many of my colleagues, I disappeared from the office one day a week to write papers and do scholarly work. My department chair, on the other hand, expected me to be available to the students during all regular university working hours. Unbeknownst to me, my regular absences were interpreted by him to mean that I was not fully committed to my job. It was not until many years later that I found out how much my absences concerned him. Once I was able to explain my strategy and demonstrate to him that the students were being well served, even when I was not available, his perspective changed. I realized then that we both had had clear expectations of performance and organizational support. But neither of us had made those expectations explicit nor communicated our needs and solutions to the other.

A lateral job change involves a different type of transition. It usually is a result of a very intentional choice rather than an expected move up the ladder. What are some experiences of a lateral move? An engineer who recently returned to a technical job shares her reasons for making this change.

I recently started a new job as a design engineer with a company that I had worked for in the past. Though the move was lateral, the responsibilities in my new job were quite different. My previous position was as a marketing coordinator, and when I accepted it, I had decided that a move out of the technical engineering world and into a world full of external interaction with clients and other professionals would be good for my personal and professional development. I jumped head first into this new opportunity. I attended Faculty of Extension courses on marketing and business management and earned a certificate. I joined marketing groups and attended seminars. The information that I acquired was inspiring and helpful, but I was not feeling like I was adding value—personally and professionally. My supervisor was passionate about "adding value" and "reaching beyond your potential," and this I also found very inspiring; I felt I needed to succeed in my marketing and business development role in order to advance to the next level of my career. The problem was that I didn't find the work rewarding.

Frustration with my perceived performance and day-to-day duties finally led me to reassess my choices. I started to daydream about doing engineering and how things were at my previous place of employment. I applied for several engineering positions and finally ended up with one that is perfect for me: I am doing technical work and am enjoying it. I came to realize that my potential could best be achieved by working in a more technical position and building on that to develop my career. The external interactions that I have with consultants at work as well as those I've created by joining an organization promoting women in science and by being more aware of opportunities around me have filled the gap that I was concerned about in the past. My transition involved realizing that there are many ways to grow in my career—without making monumental changes—and that taking a step backward to move forward is sometimes the right thing to do.

One Kind of Work or Lifestyle to Another

Women may choose to move from research to administration, from full-time to part-time work, or back into the workforce after an absence. The range and nature of their opportunities will vary with the individuals and their circumstances and culture, as will be their choices.

I recently had interesting conversations with two competent women who expressed their frustrations with the narrow focus required for success in scientific research and who sought ways to be more generalists. One of them indicated that she didn't want to read the same immunology journals all her life. She moved out of research into a senior administrative position, as associate scientific director of a research organization. The other is considering ways in which she might take on different responsibilities in an academic institution. Another friend of mine has unintentionally established a pattern of changing her work every five years. She began as a research chemist in industry, moved to analytical chemistry with a government agency, did market analysis for technological inventions, and then spent a year or so improving her skills in a second language so that she could eventually work as a freelance translator of technical material. She believes that her education and experience in science were important in enabling her to be successful in everything she has done.

I too have a bias toward nontraditional careers, especially those in which someone works out for herself just what she wants to do, perhaps by recognizing that her interests are, in fact, in constant flux, or that she is only gradually—through reflection and/or trial and error—becoming aware of her preferences. I began as a full-time research scientist but realized that I was unhappy in that role. I have now supported myself comfortably for 14 years with a combination of part-time and contract positions at a university, including administration in a research laboratory and in an organization encouraging young women to consider careers in science or engineering and by teaching courses in women's studies and in science for the general public. Sometimes this is described as "cobbling together enough work" and I sense that it is seen as second best. I, however, prefer this type of career to having one position for a lifetime. It gives me the variety I value.

It would, no doubt, be very difficult for this woman to return to a full-time senior research position, just as it would be for anyone after an extended break from paid employment. However, the realities of such deviations from the traditional career path are beginning to be acknowledged and addressed. While choosing science as a second career is still quite rare, there are initiatives to encourage women to reenter the field after working in a different area or after absences for family reasons, such as those sponsored by the Daphne Jackson Trust (U.K.) and the National Institutes of Health (U.S.) (American Association for Women in Science, 2000).

Other career changes that are more common and widely supported in the North American culture include making the transition from pure science to patent work (with or without earning a degree in law), or becoming a science librarian or a science administrator. Depending on one's definition of the boundaries of "science," these individuals may be seen as leaving or remaining within the scientific world.

In general, where training or qualifications are required, no credit is given for life experience. Strongly motivated and disciplined women in their 40s and 50s are, however, completing degrees in engineering and medicine. We believe that mature women have much to offer the scientific enterprise and encourage those who have done so to share their experiences as widely as possible.

Retirement

It may seem strange to include a paragraph on retirement in a book on success strategies for women in science. But we believe that it is an important topic to think about because making a smooth transition to retirement is a satisfying way to end a career in the sciences. A university professor of medical microbiology shares her thoughts:

> Surely the perception of a smooth departure from the workforce ranks high among the determinants of a satisfactory retirement and the sense of a life well lived. One is commonly advised to plan retirement years in advance. However, drawbacks accompany such a plan. First, if you are engrossed in your work, the future is not a particularly interesting subject—it's for later. If work has occupied a major part of your adult life, a second drawback is that one's plans for it might be outdated. A couple of years before mandatory retirement, I made lists of old and potentially new interests that I might include in my future. However, memories of pursuits that I enjoyed as a young woman (and held dearly) did not necessarily interest the mature (nay elderly) woman that I had become. I figured that I would have to rein- vent my life, with a last chance "to get it right!"
>
> It is not easy. To start with, two of the most important facts that I needed are unanswerable: "How long will I live?" and "How healthy will I be?" I took the optimistic/default view that a long, relatively healthy life awaits and will adjust as required. A workplace seminar on retirement issues taken the year before mandatory retirement focused only on financial matters! It skirted the non– "black and white" issues, namely, "How might I maintain my body, my mind, and a social existence to make for a good 'last act'?" So I decided I just would have to "play it by ear." Luck was on my side. I retired on June 30 and was reemployed the next day. The luck was not on putting off retirement, although it did do that, but in the definition of this transition period. I experienced three changes in the terms of my employment: from full time (plus) to half time, from tenure to a limited period of one year, and, most importantly, a change in tasks from research to administration. By diverting my focus to another facet of university life, I recognized that my dedication to my career had probably been a bit excessive. And I had some time to tinker with former personal inter- ests and "try out" some new ones. The year was interesting, on both profes- sional and personal fronts. When it was over, my mood was neutral—I was neither sad nor glad to leave the life I had loved. It was simply the right time for this change.
>
> I have now been "out of work" for over six months; my new life is "under con- struction" but by no means complete. For me, moving into a different position helped loosen the ties that bound me to my career. A reduced workload at the same job would not have had this result nor have been such help.

Epilogue

All the stories in this chapter have positive outcomes. Even if the transitions were painful or much courage was required, in the end, the women were happier in their work than previously. The woman who returned to her sci- entific work after enjoying being with her young child does not regret her decision; the woman who realized that her work situation was negatively affecting her relationship with her children and moved from research into teaching now loves her new responsibilities. The woman who took on senior responsibilities directing technology within her local school board learned she had both the educational background and the leadership qualities that enabled her to succeed.

Transitions, rough or smooth, lead to growth. We hope that the experiences shared by the women in this chapter encourage you during your times of change and transition and inspire you with courage to pursue your dreams.

> *By now it's clear: yours is not a time of transition—*
>   *It is a life of transitions.*
> *One change will follow another as long as you live,*
> *and some will not wait to follow—they will perch one on top of another.*
> *You will learn as you go, and you will change with the changes.*
> *You will find good company along the way,*
> *because, however unique your changes are,*
> *change is not unique to you.*
>
> J. E. Miller (1997:63)

ACKNOWLEDGMENTS

We express our sincere thanks to the women who shared their stories with us: Karina Bodo, Lisa Carter, Edna Dach, Mary Fairhurst, Suzanne Kresta, Wendy Lam, Penny LeCouteur, Jennifer Maler, Komali Naidoo, Aline Rinfret, Janet Robertson, and Patricia Schumann.

REFERENCES

American Association for the Advancement of Science. "Grants for Women in Science." *Science Next Wave* (2000) [Online]. Available at: http://nextwave. sciencemag.org/cgi/content/full/2000/05/31/12 (accessed August 20, 2005).

Bridges, W. *Managing Transitions. Making the Most of Change*. Reading, MA: Addison-Wesley, 1991.

Bridges, W. *The Way of Transition. Embracing Life's Most Difficult Moments*. Cambridge, MA: Perseus Publishing, 2001.

Daphne Jackson Trust. [Online]. Available at: http://www.daphnejackson.org (accessed August 20, 2005).

Miller, J. E. *Welcoming Change. Discovering Hope in Life's Transitions*. Minneapolis: Augsburg Fortress, 1997.

National Institutes of Health Research, Department of Health and Human Services. *Mentored Research Scientist Development Award (K01)* (1999) [Online].

Available at: http://grants.nih.gov/grants/guide/pa-files/PA-00-019.html (accessed August 20, 2005).

National Institutes of Health Research, Department of Health and Human Services. *Supplements to Promote Reentry into Biomedical and Behavioral Research Careers* (1999) [Online]. Available at: http://grants.nih.gov/grants/guide/pa-files/PA-99-106.html (accessed August 20, 2005).

INDEX